Practical Alchemy

A Memoir

Practical Alchemy

A Memoir

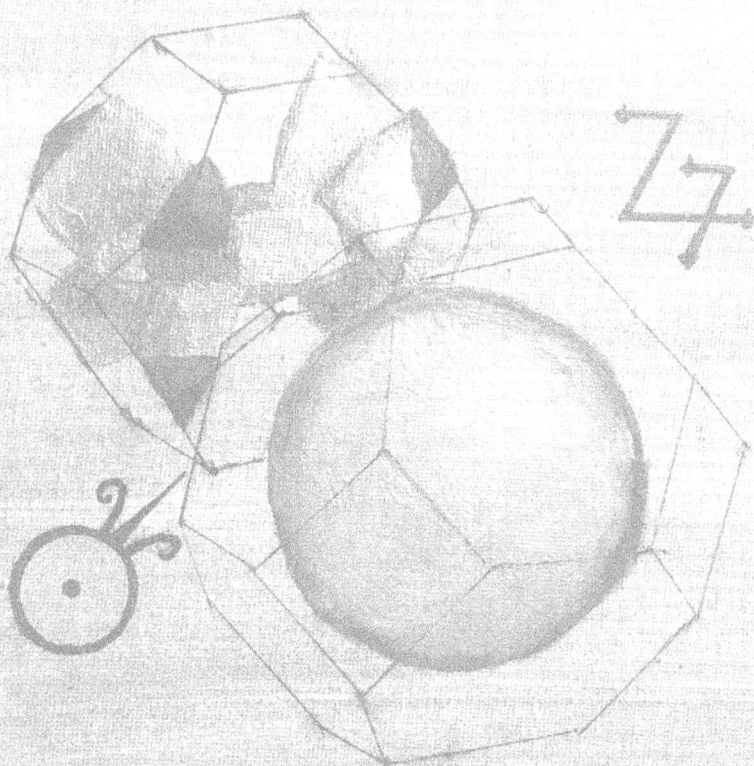

Walter A Harrison

Stanford University, USA

World Scientific

NEW JERSEY · LONDON · SINGAPORE · BEIJING · SHANGHAI · HONG KONG · TAIPEI · CHENNAI

Published by

World Scientific Publishing Co. Pte. Ltd.

5 Toh Tuck Link, Singapore 596224

USA office: 27 Warren Street, Suite 401-402, Hackensack, NJ 07601

UK office: 57 Shelton Street, Covent Garden, London WC2H 9HE

British Library Cataloguing-in-Publication Data
A catalogue record for this book is available from the British Library.

PRACTICAL ALCHEMY
A Memoir

ISBN 978-981-3277-57-1
ISBN 978-981-3278-31-8 (pbk)

For any available supplementary material, please visit
https://www.worldscientific.com/worldscibooks/10.1142/11193#t=suppl

PREFACE

The question comes up, why this title for the book. Alchemy is of course the ancient effort to convert one material into another, including the heavy metal lead into gold. It seemed conceivable then because atomic nuclei which are the distinguishing features of different elements were not known. But now we know that you go from one element to the next by changing the number of protons in the nucleus, and the electrons just follow suit. I used the title *Theoretical Alchemy* for my last book because alchemy provided a simple way to understand different compounds. The electronic structure of a pure element is simple and if you conceptually move protons from one nucleus to the next (that's theoretical alchemy) you obtain a new compound, with essentially the same electronic structure. The present book is a memoir of my life, and the various transformations that have made it a golden experience.

We may think of our own lives as a long line of practical decisions: growing up, education, choosing an occupation, getting married, having a family, and retiring if lucky enough to live that long. But out of those decisions grow the transformations which make us who we are, in my case a physicist, professor, husband, father, scoutmaster, von Humboldt scholar, and sometimes musician. I didn't seek the spotlight, but seemed often to end up in it and I'm very happy with my life as it has worked out.

I hadn't thought about writing a memoir until I read one by our friend and contemporary, Ivar Giaever. I realized that our sons would be interested and thought someone else might be. Writing a memoir has been an interesting experience. It provided me an opportunity to think back over my entire past, and as I sent successive chapters to our sons, it gave them the same opportunity. It also gave them a chance to correct some things that I misremembered. Bill and Bob were particularly helpful and at places suggested rewritings which were major improvements.

CONTENTS

Chapter I

GROWING UP IN TOLEDO

I was born to Charles and Gertrude Ashley Harrison on Long Island, New York, on April 26, 1930. In 1935, Cities Services Corporation transferred my father to Toledo, Ohio, into the management of Toledo Edison Company. He ended up being the vice president for sales, which he thought was funny. He considered himself an engineer and said he couldn't sell a gold brick. With my brother, Chuck, three years older than me, we moved to 2819 Inwood Drive in a nice middle-class neighborhood in Toledo. It was a comfortable place and time to grow up, despite the looming war.

There were three other boys my age in the neighborhood, Nick Huffer, John McWilliam, and Red Wells,

With my brother, Chuck, taller on the right, in front of our Toledo house.

and also Ray Cannon who was one year older. We played "kick the can" and other games in the neighborhood and climbed an old apple tree in a field between our houses and a large park. I remember once at dinner telling about jumping from branch to branch in the old apple tree and my mother said to my father: "Charlie, maybe you should go look at that tree and see if it's safe." My father said: "No, I don't think so. I can't control what our boys do and they have to take their chances of survival like I did." Survival?! I continued to climb the old apple tree but was considerably more cautious after that. Red Wells' father owned three roast-beef shops in Toledo and Red seemed rougher than the rest of us but for several years he was my principal playmate. We also had a basketball half-court in that field when we were older. It was an easy walk to McKinley Grade School and we all came home for lunch.

At some point I wanted a chemistry set and my father suggested that if I wanted to learn chemistry then cooking was a good way to start. I got out the Betty Brown Cook Book and found a recipe for bread. It was interesting, like a bakery. Up to that point my mother had only ever made rolls to go with dinner. My bread was quite a hit, and my mother started making bread too, but as time went on her loaves got smaller and smaller until it was back to usual. At Christmas, I got my chemistry set. It didn't turn out to be very interesting though. The only experiment I remember was one in which you mixed two clear liquids and the combination turned blue. I think one was a nickel-ammonium sulphate solution and the other potassium permanganate solution. The chemicals that came in the set were selected so that nothing could explode or do anything dramatic. My father did add nitric acid, which allowed interesting experiments with copper metal, and he added a Bunsen burner. I bought glass tubing and could blow small glass bubbles and pull small tubes. It was an activity in the basement, and I was involved in such an activity on a Sunday, December 7, 1941, when we learned that we were at war with Japan. I didn't have a clear idea of what that would mean to us so it was unsettling. Soon we listened to a speech on the radio by Winston Churchill who said that now that the United States was in the war, it was certain that our side would be victorious. He sounded very convincing to me and I never again worried about the outcome of the war.

We spent much of the summers at Sages Grove. Cities Service had set up an interurban trolley line between Toledo and Cleveland, running near the shore of Lake Erie. When arranging the right-of-way along the highway at some point they acquired a whole farm on the lakeshore and extending some 3000 feet along the shore. It was midway between Toledo and Cleveland and ended up being owned by Toledo Edison when Cities Service was dismantled by the trustbusters. When the line finally closed down, the trolleys were left in two lines near the beach and various executives had taken over the trolleys as summer cottages in what was called Sages Grove or, sometimes, Toonerville after a Toonerville-Trolley cartoon. Soon after we arrived in Toledo my father acquired one, already set up by a Mr. Jorgenson, with a kitchen at one end, a toilet in the other, and sleeping and living quarters in between. Another was occupied by a different Toledo Edison vice president, Harry Kerr, with his wife Helen and their two daughters, Mary and Barbara.

The young crowd at Sages Grove. I'm on the left, then John Emery, my age from Sandusky, and Ginger Luscombe from Vermilion, Ohio. Barbara Kerr and Dick Loomis in the back. Then Peg Emery and my brother Chuck, and Mary Kerr on the right.

Barbara and I had parallel trajectories: sharing the same class in high school and then Cornell, working at General Electric in Schenectady and later living in Palo Alto. It was not easy for friends to understand our connection and we found it best to say that we were cousins, which was essentially true.

There were other families at Sages, the Emerys from Sandusky and the Luscombes from Vermilion, and we all grew up together. We had beach fires each night, until we of the younger set went up to the trolleys to play cards at night. Our parents couldn't understand why we did not stay on the beach, but in later years they stayed up to play canasta and we had the fires and drank beer on the beach. Also in the group was our cousin Dick Loomis who lived in Cleveland and often took the bus out to Sages. He was perhaps the most intellectual of the group and was to reappear in our lives in later years.

An occurrence in grade school troubles me even now when I think about it. When I was coming home for lunch, I often saw a student named Bob Bauman who had a particularly prim way of walking that somehow annoyed me. I would see him day after day and it got on my nerves until I attacked him sometime on the way home. It wasn't much but it upset him and he went home crying. I couldn't imagine my doing it then and I can't imagine it now. I never acted the bully before, and never since. I don't remember apologizing to him, but I should have.

When I was in sixth or seventh grade, my grandfather, the Reverend Walter Herod Ashley, moved in with us. He had retired from being a Congregational minister all around the country. He seemed to favor his oldest daughter, Kate, the stepmother of Richard Loomis in Cleveland, but she felt she was too busy to look out for her father and he came to us. He seemed to be quite respected, but I couldn't detect any personality. He once noticed a crust I'd left on my plate and said: "When I was a lad, I always looked for the crust." I thought he must not be so bright if he didn't know where it was. He'd listen for the newspaper delivery every day and go pick it up, sit in a lawn chair in front of the house with the paper in his lap, and go to sleep. Eventually my father had two papers delivered, one to read and one for grandpa's lap.

When I turned twelve I joined the boy scouts. My brother Chuck was already an active scout in the same troop in the Congregational Church near downtown Toledo. Nick and John McWilliam were older than me by a month or two and got a head start, so I was tenderfoot when they were second class, and second class when they were third, and so on. I worked very hard to catch up, and finally did, reaching Eagle Scout, and even a bronze palm, in the minimum time allowed, and before they did. The scouting experience also included two weeks each summer at a boy-scout camp outside the city. The scoutmaster was Harold Wood, who worked for Toledo Edison, which might be why we ended up in that troop. We knew him also from Sages, where he reserved one of the transient trolleys each summer. Each year he organized a ten-mile hike to Milan, Ohio, ten miles away. In high school, I was a pallbearer at his funeral.

All of us in our neighborhood worked. Initially mowing lawns, later around Toledo Hospital, not far away. When we were fifteen, Nick and I got summer jobs in the Swartzbaugh Manufacturing Company, probably because my father was a friend of the Swartzbaugh brothers. I was part of a four-man team spot-welding legs on electric heaters. One would put the heater body on a rotating rig, then next put two metal legs in the rig and then the leading man, the welder, would spot-weld them together. The fourth man would stack the result on a cart. The welder was Polish, called me Professor because I wore glasses, and didn't much like me. He would occasionally remark that if he caught me outside the factory he'd beat the daylights out of me. I didn't worry too much; I thought I could outrun him. Another summer I worked in the transformer shop of the Toledo Edison Company, arranged with the help of my father. Another year in the battery laboratory of the Electric Autolite Company.

After grade school we went to Devilbiss High School, somewhat further away than grade school, but perhaps a twenty-minute walk. By this time my brother Chuck had finished high school in three years, in order to have a year in college before being drafted for the war. He had a good year at Purdue and joined Phi Gamma Delta fraternity, as my father had. I made a visit for one weekend and got a small taste of college and fraternity life. I even remember the train ride there. I could hear a "click-click-click" of the wheels going from one rail to the next. It occurred to me that I could tell the speed of the train if I knew the length of individual rails, which I didn't. At the next station I saw an old man with a railroad cap and stepped off to ask him. He looked at me as if I were crazy for asking such a silly question about something everyone knows: "Thirty two feet!". So I figured if I counted the clicks for $3600 \times 32/5280 = 22$ seconds, that would be the speed in miles per hour. I think they now weld the rails together and they don't click.

At the end of the year, Chuck enlisted in the navy and went off for training. The war was over within a year or so and he got assigned to a navy discharge station in Toledo and moved back home. I think it was arranged by a navy friend of his, Jim Vaughn, who had managed to get both of them transferred there. Jim moved in with us also and was another older brother to me. They were together with friends on New Year's Eve and Chuck had too much to drink. He took off from the party in our family Lasalle and had a big accident, from which an older woman in another car died. I never heard much about the details nor about his condition, only from Jim. The Lasalle lost the front left fender and was out of commission for a year or so because no car parts were available because of the war. That left us only with my father's company car. I eventually hunted up a fender from a junk yard that was almost like the original and the car returned to service. I was mobile again.

A number of our friends signed up with the Marine Reserve in high school. They thought it was a big deal. It paid $150 per month, about the same that Nick and I made working full time at Swartzbaugh. They went to an evening meeting every week and spent two weeks in the summer at Camp Lejeune, North Carolina. They pointed out that if there were a war, we'd all get drafted in any case and they would be in a better position. I asked my father about it, and he said no way! "It's best to stay as far removed from the military as possible." He turned out to be absolutely right. When the Korean War came all these friends went off to Korea and I continued merrily along in college. Fortunately, they all returned alive.

Ray Cannon had joined the high school fraternity, Phi Chi, the year before we went to high school, and Nick, John and I pledged our freshman year. Red Wells had chosen not to go to Devilbiss, but to a vocational high school and we saw little of him. These fraternities, and the corresponding high school sororities, arranged dances with live orchestras, selling tickets to support them, and dates became central activities. Older fraternity brothers usually provided transportation and two or three parents served as chaperones at each dance. Toledo high-school dancing consisted of walking, side-by-side with arms around waists, around the floor. My father didn't consider that dancing, and we all went to occasional dancing schools with real dancing. Once a pair were "going steady", they were "pinned", and she wore the fraternity pin. I managed that about four times during my high school. I lost my pin sometime early and the replacement was notable. My friend John Browning's sister had been pinned to a Phi Chi earlier on, Bob Chapius. She ended up with the pin, which became mine. By then Bob Chapius was a local hero, the leading running back for the University of Michigan football team.

Some of us joined school athletic teams, John McWilliam played basketball and I ran track. I was part of the mile relay and one year, we even got a silver medal at the Ohio-State championships in Columbus Ohio. During the season there were regular meets with other high schools in Toledo and big meets with many schools participating, at various colleges around the state, so it became a major activity. I had begun smoking cigarettes but abstained during the track season. In my senior year I was captain of the team and won a first in the all-city meet in the hundred-yard dash. I was referred to as a "scholarly speedster" in the Toledo Blade newspaper.

Some of our other activities were less widely recognized. The minimum age for drinking alcohol in Ohio was eighteen for beer with less than 3.2% alcohol and twenty-one for everything else. However, we all had IDs which could get us drinks in bars which were not too choosy. One way was a birth date, typed on tissue paper, moistened and stuck on top of the real date on a driver's license. I expect one could have blown it off in case of a raid, but it never came up. Looking for bars which were not too choosy took us to many corners of the city which one might not ordinarily frequent, and I have often wondered how much that had impacted my life. One place I remember, I think called Monaki's, was in the Czechoslovakian neighborhood across the Maumee River, the East Side of Toledo. It was small and dark and the prevailing music from the jukebox was near-Eastern European violin and accordion and most soothing along with a glass of draught beer. It may have been responsible for a part of my yen to spend some time in Eastern Europe. Another place was Nelly's Tavern, outside Toledo, with country music on the jukebox and square dancing with live music on some weekends. That's a different feeling but I still enjoy the country music that came before country rock and I still enjoy having a glass of beer in a comfortable bar and getting acquainted with the locals along the bar. I think of one night more recently when I noted that I was friends with some fifteen of the twenty people sitting at the bar in our current local, Antonio's Nut House, but that's a much larger fraction than usual. My wife Lucky and I are somewhat special in Antonio's; it is unusual for a *couple* to frequent such establishments. Antonio provides free peanuts and the bar is known for the peanut shells covering the floor.

My painting of Antonio's Nut House. From left, Siegie, Mark, Charles, Dale, Lucky, Kenny. Kelly is serving.

High school was my first chance to take a foreign language. I was intent on taking German and surprisingly enough it was still possible in the midst of World War II. Only two years were available, rather than the traditional four, and they were taught by Frau Gerding, raised in the United States but *very* German. We were thoroughly trained in grammar and by second year, read novels which stick with me still. It was a real contrast to take French my second two years with Pierre Pasquier. He was delightful but the language training was *comme ci, comme ça.* He read to us from *Cyrano de Bergerac* and brought us, as well as himself, near to tears in the sad parts.

Some of the other teachers at Devilbiss were quite interesting. There was Clyde Kiker, who taught economics and sociology. He was rumored to be a communist, and in those days that rumor could have cost him his job, but he was also the lawyer for the teachers' union. We thought that was what saved him. I signed up for his sociology class and liked it. It wasn't very taxing; he didn't have much use for the official textbook and assigned other interesting reading and each student gave a talk during the semester. I gave one on a book about socialized public utilities, which pleased him initially as I recounted the benefits. However, my father had strong feelings the other way and a lot of back-up information. We had gotten that all organized so in the second half of my talk I presented the other side. At the end of my talk Kiker stood up with a big smile and said: "Well, we're going to take those arguments apart piece by piece!" Actually, there wasn't much time and that was the end of it. We continued to interact well for the rest of the course and he gave me an A. Once after I had graduated, I returned and sat in on a class. He was talking about how the United States and Britain exploited the Chinese because they could get away with it. One student spoke out, which was usual in the class, and said: "Well, whatever they can get from those Chinks is OK with me." Kiker responded: "Thank you for the punctuation!" Getting an A was not unusual for me. I got A in every class but one during my time in high school. That one was an English class with Miss Warner. I *did* disagree with her at times but I think she gave me a B for sometimes whispering to a student next to me when she was talking. I like to think of the B as the false stitch which each Navajo women is said to put in every rug, with the thought that death would come from a perfect rug.

Miss Sampson was a math teacher whom I liked, teaching solid geometry. One time when I was going through theorems in the text, I noted a theorem that didn't look right to me. I can't remember exactly, but it had to do with a point on a line and a parallel line in a plane and asserted that the point must be on the plane. I could see a circumstance where it wasn't. So, I wrote a description of that situation and wrote two careful proofs, following all the proper procedures, and using the faulty theorem. One proved the point was on the line and the other that it was *not* on the line. I took them to Miss Sampson and said I bet she couldn't find an error in either proof. She didn't see anything in a brief look and said she'd look at it at lunch time. I came back in the afternoon but she hadn't yet had a chance to look at it, though I learned later that she had discussed it with another teacher at lunch. In the end she solved the problem. She found in *another* solid-geometry book the same theorem, but with a qualifying clause that ruled out my circumstance. I heard that after that episode, every year she told her class to check if this theorem is corrected in their book, and how a student had found the error.

I had a similar occurrence with our physics teacher too, Mr. Archambo. There was a picture in our text of a bathtub with water in it, and a bathmat on the side, dipping into the water inside and dripping onto the floor outside. The problem was that the dripping edge of the mat was higher than the surface of the water. I *knew* that could not happen so I showed it to Mr. Archambo and said "Look, they got the picture wrong in the book!" He said: "Oh no, it's correct, water'll soak up there and drip off as they show." I argued: not if it's above the water level; otherwise you could catch it with a little water wheel and get free power as it ran back into the tub, a *perpetual motion machine*. He said, "OK, you could, but you wouldn't get enough power to be worth the trouble." This time I didn't push it. He was a practical man, even if he didn't understand physics very well. I was quite proud of myself for this argument and at the time didn't even know about thermodynamics, in which context the argument would have been obvious and immediate. Some time later Mr. Archambo taught me the elements of differential calculus, which was completely new to me then, but second nature now. It was also during that class that I realized that my future was in physics, not in engineering for which I had been applying to colleges.

I entered the competition for the Westinghouse Science Talent search during this senior year. It involved a written exam and we needed to turn in a report on a research project. I decided to try to understand a familiar gadget, in which a row of steel balls are hung touching each other. When you pulled out the first and let it go, the ball at the far end sprang out with all the others still. I couldn't work with all the balls at once so I pretended there was a tiny space between each one. I could work out the collision with the first ball, and then I could work out the collision of the first with the second, etc. I was conserving momentum and energy and could only make sense of it if I said momentum going the opposite direction was considered as negative, a new concept to me. I told Mr. Archambo that and he said: "No, if you've got momentum, you've got momentum; it can't be negative!" I wrote it up as something I'd discovered. When I got to college, in the first few weeks of physics they talked about positive and negative momentum as if everyone knew that, everybody but Mr. Archambo. My write-up may have puzzled the readers, but they must have been sympathetic because I ended up with an "honorable mention" in the contest.

For our Senior Play we put on *You Can't Take it with You*. Nick was the wise grandfather, I was the colored servant, and another friend, John Browning, played the Cary Grant part. On Saturday we had a matinee and an evening performance. The three of us, and a couple of others, then went out for dinner and stopped at Nelly's Tavern. It shook Nelly up at first because I still had my black-face but she adjusted. The problem came that night when my parents went to the evening performance and overheard a student tell someone that I was drunk. It wasn't true and all went well at the play, so it wasn't a big problem. My parents were always accepting, though not always enthusiastic about all my activities. My father, in particular, objected to "that screwball" Paul O'Connor, who was one of my closest buddies toward the end of high school. He ended up marrying a girl I had dated, becoming a Deputy Sheriff in our county, and finally retiring in Florida.

When it came time to apply for college admission, I applied to MIT, thinking it was where I would go. I don't know how much of that decision came because of my father, how much from school, or from a neighborhood friend, Don Robertson, a couple of years older, going there. I also applied to Cornell, Williams College and some others. In spring I took the train east to visit MIT, stayed overnight with an aunt and uncle in Larchmont, north of New York City, and then took the train to Cambridge. It was a rainy Sunday when I arrived at the Phi Gamma Delta house at MIT, where Don Robertson was living. There were liquor bottles on the floor, as I remember, and some members were moving to their desks to study. It looked like kind of a grim life, but I thought maybe that's what college life was. They arranged a date for me, whom I didn't particularly like, and the next day I went to MIT to talk to people there. It wasn't so jolly heading back home but I stopped at Cornell on the way, and also stayed there in the Phi-Gam house, with Jon Ayers, a friend from Toledo. That morning it was a beautiful day as I headed across the campus. It was during classes, but a student walking along near me burst into song! I stopped in my tracks, looked around, and decided I was not going to MIT. I'm going to Cornell! I remember telling my mother about my decision, and she said: "Well, wherever you decide to go I'm sure you'll feel it was the right decision." I didn't believe the statement, but I certainly *did* end up feeling that this was the right decision for me at the end of my high school years.

Chapter II

UNDERGRAD AT CORNELL

I went ahead and made all the necessary arrangements for going to Cornell, even received a nice scholarship that paid tuition for a while. I signed up for a dormitory room, for Freshman Camp, and orientation. Then, just at the end of the summer, I got very sick, with a perforated appendix. Fortunately, antibiotics were just coming on the market and I survived a life-threatening situation. However, I missed Freshman Camp and orientation and the first couple of weeks of school, which included fraternity rushing. When I finally was able to take the train east (and even had a Pullman bed), I had lots of catching up to do. My roommate, Art Stein, was helpful. We had a room in temporary dormitories just below the library slope, a reasonable walk from classes.

When I had applied to Cornell, it was to Mechanical Engineering, which was where I was admitted. I tried on arrival to switch to Engineering Physics, a new department, in its third year. That wasn't allowed so I proceeded with Mechanical Engineering. The first class I walked into was calculus, taught by Mark Kac, an interesting Polish mathematician who had written the text with a Professor Randolph. When he spotted me sitting there, he said: "Who are you?" I told him and the class burst out laughing. I thought about walking back out. "Ah, the mysterious Mr. Harrison. Sit down and be confused." It turned out that he had been taking roll each class and asking if the mysterious

Mr. Harrison had arrived. In the end I thought he was a wonderful teacher could not write a better text. We asked him much later how such a wonderful teacher could write such a terrible text. He said, "What? And do myself out of a job?" Once he assigned a numerical problem and a student asked him what value to use for *e*. He said it's in the book but the student said no. Before the next class Professor Kac was in front, leafing through the book. When the bell rang he just said "That damn Randolph!".

I had contacted Jon Ayers, and had dinner at the Phi Gamma Delta house, and also Jim Vaughn, who had gone back to Cornell after being discharged from the navy, in the Phi Delta Theta house. After a week of rushing at the beginning of the year no one was interested in thinking about new members so that was the end of that. In my mechanical engineering classes was a David Plant. He was from Toledo and I had met him at an interview for Cornell scholarships in Cleveland the previous year. He had joined Delta Upsilon (DU) and took me to dinner there. I had limited interest because Chuck had given me a list of fraternities with his evaluation. He had listed DU as OK but heavy on athletes, a jock house. However, I very much liked the people I met there and in the end they offered to take me and I accepted. That was another decision which I have never in the slightest regretted.

Fraternities at that time were very much different from before, or after. The older members were almost all veterans from the War. They were interested in getting an education, and enjoying the process, but not in the antics that one usually associates with fraternities. As pledges to the fraternity we were required to work at the house part of the weekend. I was happy with that, and then didn't have to do it in later years. The second year I moved into the DU house and it provided a very pleasant and congenial place to live and study.

We had a bar in the basement and most weekends brought in kegs of beer and had a nice party, open to friends and anyone else who wandered in. Half the male undergraduates were in fraternities and they seemed to be the center of social activity at Cornell.

Each year we had a few house parties for which we vacated the second floor for the visiting girls to stay. As far as I know, no one violated that rule. We had an outdoor porch on the third floor with a row of beds.

That's where I slept, even in the cold Ithaca winters, but I did have an electric blanket. We ran the house, hired the cook and cleaning man. In my fourth year I was president. By then I could see that things were changing. The younger brothers were yearning to lean on the pledges a little, but as long I was there it was under control.

In the second semester of my freshman year I was able to transfer to Engineering Physics. The plans for the department fit exactly with my interests. It was a five-year program, as at the other engineering schools that time. We took as much physics as the physics majors, but also took the full set of engineering courses: metallurgy, casting and pattern making, welding, mechanical drawing, etc. There were about twenty of us in the class. Each year a few transferred out to less stringent programs and a few transferred in, so the total remained about the same. The original idea was a complete program to train students to be research and development engineers in industry, which sounded good to me. It didn't really work out that way; in the first few years at least, almost all the students went on to graduate school, and a good fraction of them to do theory. I expect it was because technically capable high school students were encouraged to take up engineering in college, but once they got there they decided physics was more appropriate. This was an easy way to adapt within the Engineering College. We carried a heavier course load than those in humanities and sciences, so we spent much of our time on class work, but I think I managed to enjoy the full college life in the university I came to love.

Before I left for Cornell, Chuck said I should take up skiing; everybody skis at colleges in the East. We had done some skiing in Toledo but it was with a strap on the skis over our galoshes and not real skiing. I went to a ski-club meeting where an older Norwegian student talked about what equipment to buy. He said the boots didn't matter but you needed very good skis. I thought it was silly advice but I did buy cheap army ski boots, *and* skis. There was a local Cornell ski hill which I tried, but later I went with others in DU to real skiing, at Mont Tremblant in Canada, for one trip at least. By then I had real ski boots.

With a ratio of three men to each woman at Cornell, dating was a challenge. Barbara Kerr had come from Toledo at the same time, and joined the Delta Gamma sorority. She arranged some dates for me, and friends in the fraternity helped each other out. At one point I even went to some youth programs at the Congregational Church in town, but that didn't last.

In my junior year, Nels Schaenen, an older brother in the house, encouraged me to run for class president. I took up the challenge and very much enjoyed running. He arranged for me to have dinner in many other fraternity houses to meet my fellow juniors. It succeeded and I became President of the Junior Class. That also put me on the Student Council, which was a much more active group than the class council. It greatly broadened my contacts among the student body, including of course many women. At the end of the year my friend Jim Gibbs was elected President of the Senior Class. Jim was somewhat rare at Cornell in those days, a black student. He had grown up in Ithaca and was a fine young man. He was to end up as a Professor of Anthropology at Stanford. He, his wife Jewelle, and their two sons remained good friends thereafter.

During my junior year I dated a Kay Kirk who was involved with the CURW, Cornell United Religious Work, and from Saint Louis. I visited her family once and she visited mine once. I expected my mother to like her very much. My mother being a PK (Preacher's Kid), I thought they had common background and would hit it off well, but it didn't seem that way. Kay graduated from Cornell at the end of the year and took up a job in Colorado. She ended up marrying a rancher from Minnesota; it is hard to imagine her on a horse. The next year I dated Joan Sanford, at Elmira College. By then I had a car and it was convenient. My family had decided to get a new car for my mother and I got the '47 Chevy she had been driving. Joan graduated that year and I lost track of her. I was sometimes worried about that. Lots of girls got married as they left college, but I was far from having marriage on my mind. It wasn't until my fifth year that I met Lucky Carley, whom I would end up marrying, but not right after graduation.

I remember one unhappy time during my junior year. I applied to be a counselor at the Freshman Camp the next fall, the camp that I had missed in my freshman year. Eventually I heard that I was not accepted. I was devastated because I thought of Frosh-Camp Counselor as exactly the kind of person I wanted to be. Then finally I figured maybe there was someone involved in selection who didn't like me, and I could live with that. Now it reminds me of a story I heard later: there was a very virtuous man who spent much of his life doing good deeds. Nevertheless, his personal life seemed to be one difficulty after another. After a particularly bad incident, he knelt in his back yard to pray, asking: "Dear Lord, why do you trouble my soul so?" The sky darkened, and some lightning flashed, and God replied: "Some people just piss me off!" Or an equally sacrilegious story of the same virtuous man, playing golf with a man who was his exact opposite, greedy and having all the attributes of the worst of humanity. During the game, the sky again darkened and God appeared among the clouds. A bolt of lightning hurtled down and struck the virtuous man. Then God was heard to say: "Oops, I missed!"

My summer jobs during college proved more interesting than before. The first two summers I worked at the central research laboratory for the Owens-Illinois Glass Company in Toledo. The first year had to do with an occasional problem of molten glass sticking to a metal mold used in making glasswares. Then I saw one of the machines and noticed a little pot of something brown and gooey with a long-handle brush. The operator occasionally swashed the mold with it. I heard that the origin of this procedure was from an engineer who had earlier noticed that one particular machine never got stuck. He watched that machine and noted that the operator, chewing tobacco, would occasionally spit into the mold. So the company purchased some Mail Pouch Chewing Tobacco and proceeded. I didn't advance the science of glass making much further that summer.

The next summer was better. The company sought a way to measure the thickness of thin films adsorbed on glass. There was an approach based upon the ellipticity of light reflected from the surface at the polarizing angle. My project was to find a way to make the measurement. I was quite proud of my approach, using a birefringent crystal, calcite, to split the reflected beam into two beams with the two directions of polarization. These would be observed through a polarizer which could be rotated until the two beams were equally bright, and the angle would tell the ellipticity. I don't know if the device was ever made. In the course of this I realized that though two crossed polarizers let almost no light pass, a third polarizer between them at 45° would allow some to pass, up to an eighth of the original beam. I'm not sure if that was known, and now I don't find it in Wikipedia.

The summer after my junior year I got a job at Oak Ridge National Laboratory in Tennessee, requiring me to get my first Q clearance. The city of Oak Ridge had only recently been opened up but the laboratory was fenced and guarded. My assignment was to an area called X10, where there was a nuclear reactor operating. My supervisor had a station where a beam of neutrons came through a small port in the side of the reactor. We had a neutron counter and could measure the absorption of a specimen by counting neutrons in the beam with, and without, the specimen in it, and the background count when the beam was turned off. It was a slow process which required someone to be there reading the meter, moving the specimen, and adjusting a device which selected the energy of the neutrons observed. For the summer that someone was me. It wasn't very challenging, but it had interesting sidelights.

I was astonished that there was complete segregation in this government laboratory, "white" bathrooms and "colored" bathrooms. I was acquainted with a black maintenance worker and could learn from his perspective on it. Then because of these discussions I had the honor of being called a "nigger-lover" by one of the white workers operating the reactor from the office next to ours. Another interesting sidelight was figuring out what would be the most efficient way to divide the time between counting the beam without the specimen, with it, and the background. It turned out that my supervisor had me spending too much time on background. He didn't appreciate my pointing that out.

I shared a room in town with another student from Cornell, Frank Loeffler. We became acquainted with a girl's softball team in town and for a while I dated one of them. Later I noticed a pretty girl sunbathing at a city pool where I was swimming. I'm not sure I would ordinarily have approached her, but I figured this is Tennessee and very friendly so I did. She was a nurse from Boston and later told me she wouldn't have responded, but thought this is Tennessee and very friendly so she did. She had a complicated schedule but on occasion I could count neutrons on a schedule to match hers and we could see the countryside. There was a pleasant tavern outside town on the road to Clinton. Later I learned that this was a dry county and it was a speak-easy, but it wasn't obvious.

Another technician friend of mine mentioned that he wanted to go up to the hills and buy some moonshine. I suggested I could drive him up and get some for myself. We did and split a gallon. I was a little nervous about how safe it was, but he sat down and got drunk when we returned. The next day he was fine so I knew it was OK. It was as clear as water and seemed terribly harsh to me, but it was a nice memento. I had it in the trunk of my car when I returned to Cornell and at that time decided to go through Canada, not far out of the way from Toledo. Only after did it occur to me that I had smuggled illegal booze into Canada and back into the States. It jolted me at first but then I realized that if I had been stopped, and was thinking, I could have said it was paint thinner. If they smelled it, they would believe me too.

One thing which remained from my being president of the junior class was being a student member of the Faculty Committee on Student Conduct, which reviewed the cases of students who got in trouble. It was moderately interesting but became dramatic at the end of my fourth year. A sizable group of my friends that I knew through the Student Council and other ways decided to do a prank during the final exam week. Most of them were seniors, like me, but not in engineering so they were graduating. I didn't know about it, perhaps from being in Elmira for a date. The prank was a bad idea and backfired. They took over the campus radio station and announced that there was a war brewing in Europe. It quickly got quite silly, like quoting the University President as recommending that students soak their feet in warm water, so everyone was expected to know it was a joke. However before that one student, who had been a refugee in Europe, made moves toward jumping out of a fifth-story window. Fortunately he was stopped and there were no major repercussions, but the whole group of my friends were brought before our committee. Also attending was the University President, Dean Waldo Mallott. He was particularly incensed because they also quoted him as saying "The only thing we have to fear is fear itself", a real quote

from Franklin Delano Roosevelt. That hit a sore spot because after
Mallott was inaugurated, a long section of his acceptance speech
appeared in the Funny Coincidence Department of the New Yorker
Magazine, along with the long, and mostly identical, quote from an
earlier speech by the president of Sarah Lawrence University. There was
a good deal of discussion about dismissing him as president for
plagiarizing his speech, but he survived. (I was an usher at his
inauguration and heard his speech. I even remembered the part he
plagiarized as being particularly good.) He didn't enjoy the reminder,
and said that that alone was enough requirement to throw the entire
bunch of students out of the university and cancelling their degrees.
Mallott left the meeting and the committee was more moderate, but I
believe their degrees were postponed and they did not go through
commencement. I've wondered if I would have had the sense to stay out
of it if I'd been around when it was planned. Maybe after serving on that
committee I would have.

After my senior year I had a job at Sandia Laboratories in Albuquerque, New Mexico. It was arranged through Dr. Sack, my advisor in
Engineering Physics at Cornell, who was a consultant there. At some
time when it came up that I was going to Albuquerque a pretty young
coed I had met said "Oh, wonderful! You can give me a ride to Oklahoma
City, which is right on the way." I had some misgivings about transporting
an unmarried women across so many state lines for one thing. I went for
advice to a middle-aged friend, E. M. Johnson I think, in the leadership
of the Telluride House, next to DU. I needed to learn how it would look
through middle-aged eyes. He thought it sounded delightful and saw no
problem. He also told me of two medical friends in Los Alamos that I
should look up. So off we went west. We stopped in Toledo en route. We
got two rooms at each motel and she paid her way. It *was* pleasant to
have company, but I admit I was happy to drop her off in Oklahoma City.

Sandia was a somewhat secret laboratory and I had no idea what
they did, though I thought it had something to do with military
electronics. When I got there I found that it was the place where nuclear
bombs were made including the two which ended the war. I stayed in a
house in Albuquerque with Ted Waddell, a DU brother from Cornell,

and a friend of his; both had been called up from army reserve and stationed at the Kirtland Air Force Base, adjacent to the Sandia Base where the Laboratory was.

I was assigned to a model test group, which was great! Every morning we'd drive jeeps and a truck out to Coyote Canyon on the base. There was a mobile home there and the first order of business was making coffee. The boss was Deacon Palmer, an easy-going westerner in a cowboy hat; we had one service man who handled high explosives, and a couple of others I don't remember. We had a forklift for moving around blocks of concrete that represented models of buildings. They also contained pressure gauges that measured the rise and fall of pressure when a shock wave came.

Occasionally we'd have a test for which boxes of high explosive were piled on a wooden tower we'd make. Then we went to sit at the top of the hills adjacent to the site and watch as the explosives man set off the charge. The tower disappeared and I believe I could see the shock wave fly out and past us, light reflected from the pressure discontinuity, but I'm not sure. Afterward, back at the base, we watched films of the wall taken at the time of the explosion. I noticed that you could see a flash of light on the wall, as the explosion went off, and then some puffs of smoke on the face of the wall, and then eventually the shock wave striking the wall. I asked what the puffs of smoke were, and Deac Palmer said they were bolts from the wooden tower, which arrived much sooner than the shock wave. They came like bullets and I realized that they were actually going in all directions, including toward the top of the canyon. I'm not sure we watched the explosion directly more than once.

The bolts weren't the only danger in Coyote Canyon. One day working in a man-hole-like pit, I watched a beautiful black spider making a web. Then I spotted a red hourglass on it and realized it was a Black Widow. I'd never seen one before but they were common over there. Another time we happened on a rattlesnake. Deac made a rig to catch him: a wire doubled over in a pipe. Holding the two ends at one end of the pipe, he put the loop at the other end around the snake's head and caught him. We finished him off and skinned him and I ended up with a fine trophy. All of this delightful activity lasted till Dr. Sack visited the lab from Cornell late in the summer. He was distressed that I had been made part of a labor team and the rest of the time I spent plotting pressure graphs at the base. Deac called them plots of noise.

Ted Waddell was going out with a beautiful Spanish girl in Santa Fe, and arranged for me a date with a friend of hers. I was delighted to have a chance to get acquainted with old Santa Fe. We occasionally visited Tiny's, on Cerillos Road south of Santa Fe. They had dance music and I remember particularly watching a very stately Spanish couple who danced beautifully. One time many years later, Lucky and I had dinner with friends in Tiny's, and I saw such a stately woman sitting at the head of a large family table. I went and asked her if it could be the same, and indeed it was! Her husband had died, and she was pleased to hear that I had so much enjoyed their dancing.

I drove up one weekend to Los Alamos. With my Sandia badge I could enter the town which was still fenced and closed to the public. Driving along a back road next to the Lab I was straining to read detailed signs about trespassing and someone coming in the other direction may have been doing the same thing. In any case we collided. We both got out, prepared to be angry, but both quickly calmed down and surveyed the damage. His was worse than mine, a tire gone flat but I just had a smashed headlight. We each apologized, he called AAA and I went on. I did look up the friends of E. M. Johnson of Telluride; they thought after my accident I could use a drink, and we did enjoy one together. When I got back to Albuquerque I went to a little car shop next to the lab. When the owner found there was no insurance coverage he said he could fix the light for $25 if I didn't need a perfect fix. I took him up on it.

I returned home at the end of the summer, very happy with the golden tan I had acquired in the canyon, and happy to drive it alone. The tan disappeared in the steaming Midwest on the way back, but it was a nice ending to the series of summer jobs.

On our fifth year at Cornell, Dave Plant and I moved into the Q&D Apartment. Quill and Dagger was a Senior honorary society of which we were both members. They had a meeting room at the top of a dormitory tower, close to where I had first stayed at Cornell. The tower room had a fireplace and balconies and tall windows. Just below it was an apartment which we were able to reserve. There was a small elevator serving just the meeting room and our apartment, and a private door to the outside. We continued to have our meals at the fraternity house, which was a short walk away. Later in the year we had occasional parties in the meeting hall, though I think it was supposed to be a secret place for members only. We'd have a fire in the fireplace and I'd serve shots of my Tennessee moonshine. Someone tossed theirs into the fire and made a nice whoosh and flare. It was then often repeated.

Liz Thomas, who had been Joan Sanford's roommate at Elmira College, had been offered a graduate fellowship at Cornell by the Ford Foundation for a one-year course leading to a Master's Degree and a New York teaching certificate. She sometimes dated a friend of mine and suggested I call another girl in the program, Lucky Carley. I did call and ask her out for a coffee, but she said she was washing her hair and was unavailable. Apparently Liz had over-advertised me to Lucky and I sounded formidable. The same thing happened the second time I called but I decided to give it one more try and the third time she agreed. We went to the Clinton House bar downtown and I asked her if she would

Spring House Party at DU in 1953. Lucky and Walt are at upper left. Next to us in back are Rit Rittershausen and Skip Byron, whom I had recruited from Toledo for DU.

like coffee or a beer. She later told me she was thinking more of a whiskey sour, but she took coffee. We saw each other after that but I was committed for the Fall House Party. With the shortage of women, one had to plan ahead. But I definitely did invite Lucky to the Spring Houseparty. She invited me down to her home on Long Island at Thanksgiving and we went to Philadelphia for a football game before driving back.

We saw each other regularly after that. She came with me by train to Toledo at Christmas. Earlier we had seen a movie with a house with a particularly glorious main staircase, and I had quipped: "That looks like our house, but we have the stone vase on the other side of the stairs." She may have been surprised at our real modest abode. We had the usual raucous New Year's Eve party with my Toledo gang. She also came at spring break. Around that time my cousin Dick Loomis left Gethsemane Monastery and came to our house. He had become Catholic and left college at Oberlin to go to John Carol, a Catholic university, and had finally gone into training at the Trappist monastery. He became so zealous on approaching Easter that the brothers feared for his survival and he was sent out. His parents were in Japan at the time so he came to us. Trappists don't speak so he had difficulty communicating; it was a strange situation for Lucky to visit. John McWilliam was married while we were there, I think the first marriage among our group.

At this stage we were well into the Korean War. I had been deferred from the draft for college, but my plan to go into industry upon graduation was not a possibility. One option which looked good was to do a year of graduate work, get a Master's Degree, and I could then enlist as a nuclear supervisor, a technical position rather than going into the trenches. I planned on that option and applied for graduate fellowships and graduate schools. One that sounded particularly appealing was going to Germany and I had heard that Göttingen was a fine technical university, much like Cornell. I applied for a Fulbright Scholarship to go there, but also applied for a general National Science Foundation Fellowship with which I could go anywhere. Late in the spring I heard that I got the NSF Fellowship. It paid better than the Fulbright, and I was beginning to think that the University of Illinois was a better idea in any case so I accepted it and withdrew my Fulbright application. That was another decision I never regretted. It's interesting that many years later our son Bill won a Fulbright fellowship and ended up in Göttingen for the first part of his Fulbright year. I think it was much better for him than it would have been for me.

I was accepted at Illinois and visited there. Professor Almy told me how it was laid out, a simple set of classes to get a Masters, essentially all of the courses needed for the PhD and if I continued, doing research for a few years, I would have a PhD. My fellowship would continue. It was a big contrast to my view of graduate work at Cornell. I pictured a long drawn-out period of experiments till one was finally tapped as finished. Now I felt I was on my way. At Cornell graduation Lucky and I dressed up in our caps and gowns, hers more elegant with the colored graduate hood. Both of our sets of parents came and we all got together for the first time.

Lucky in the mean time was offered a teaching position in Spring Valley, California, as was her roommate at Cornell, Jane Nicklen, Nicky. Lucky's father gave her an old Plymouth and the adventuresome pair headed out to the West. They stopped in central Ohio on the way and I met them there. The next day they were off, and I was getting ready to head off to Illinois.

Lucky on her way west to teach.

III

GRADUATE AT ILLINOIS

I rented a basement apartment in Urbana, a mile or so from the University, in the house of Bob and Betty Swenson. Bob was the cellist in the Walden String Quartet and Betty was a delightfully loose cannon. I was looking forward to hearing the cello upstairs, but it turned out that he never played at home. Betty was fun. I remember one afternoon she was dragging a lawn chaise lounge across the backyard to near the trash. She couldn't decide whether to throw it out so she put it *near* the trash to let the trash man decide. I was often up in their kitchen with them over a glass of sherry. I found someone to share the apartment, Ted Husek, a psychology graduate student from Chicago. We had a double bed and a single bed and I slept in the double.

I got acquainted at the DU house and had regular dinners there. I went to their parties and made good friends with a couple of the older DU's. That got me acquainted with a few undergraduate girls that I went out with. Toward the end of autumn, it was a Bobo Spencer from Chicago.

I took four courses in the Physics Department, but otherwise had little to do with the department. Three courses was the usual load but with a fellowship, rather than teaching assistantship, four were allowed. One was solid-state physics taught by Fred Seitz. He was somewhat stuffy, but I found a way to make him unwind. I went up after a class and said I had a question. He took his usual pose, ready to answer a physics question, listing references with it, as he prepared for the question. I said I was going to be flying out of Midway Airport in Chicago and where should I park? His face lit up and he enthusiastically told me not to park in the airport lot, but a block away was a little lot, and so on.

One interesting course I took was on using the Illinois computer, called the Illiac. At that time there were only a few computers of such a size in the world, but it was tiny by today's standards, much less powerful than the hand-held calculator most school children have now. We wrote programs in machine language, printed them on a teletype tape, and fed them into the machine ourselves. Still it was impressive, but when I went to GE the laboratory didn't have a computer and it was seven or eight years before I came back to it.

Lucky and I were exchanging several letters a week. We were both smoking in those days and I remember one night while I was writing to her, I started to light a cigarette, but noticed that I already had two going in the ashtray. Lucky was busy teaching first grade, and dating in California. When Christmas recess came Lucky flew home to Garden City. I took the train east to visit, and we became engaged. Lucky thought it was proper that I ask permission of her father, which I did. He said: "Well, what does Lucky say?" There was no problem.

After Christmas we went to Toledo and had our regular New Year's Eve party. My friends and I had developed a custom of putting on a play of the *Three Billy Goats Gruff* during the evening when everyone was in their cups. This involved collecting the men present, discussing for one or two minutes who would play which part, and who would be part of the bridge (four kneeling men made the bridge). The ladies were supposed to watch and it took only two or three minutes. It was a tradition that Lucky and I continued for many years at our parties, though it died out in Toledo soon after I left. That crowd then heard from me about continuing it in Schenectady and one year they decided to reinstate it in Toledo, but they did it the wrong way. They had rehearsals and singing as well as music. They were happy enough with the result, but never repeated it.

After New Year's, Lucky flew back to California and I took the train to Illinois. As luck would have it, I ran into Bobo Spencer on the train from Chicago to Champaign and she was not enthusiastic about the news of my engagement.

When I got to the house, Betty Swenson informed me that Ted Husek had found out that his girlfriend in Chicago was pregnant. They had gotten married and were now in our apartment. Betty wanted to ensure that they didn't drive me out and take over the apartment. So, I continued sleeping in the double bed while they huddled in the single. After a few nights, they found someplace else and I don't think I ever saw Ted again. I found a nice senior student in engineering whom I don't remember well, but I did like him. He had a much newer car than mine and didn't lock it when parked outside. I figured there was no need for me to lock my older car parked next to his; a thief would choose his! One morning I found my car radio had been stolen. My old Chevy radio apparently fit the thief's car. I remember in June my roommate collected all his liquor and wine together and poured them into one jug to put in the graduation punch at his fraternity. I didn't go to that party, but I'm sure it helped.

During the first year I needed to pick an advisor for my thesis work. I had pretty much decided to go with Professor Dillon Mapother in experimental low-temperature physics. Students with NSF fellowships were a priority because the professor didn't need to supply salary, so I knew I could pick whomever I liked. One day at the end of a solid-state physics class I went up and asked Fred Seitz what it was like doing a thesis in theory. He told me to come to his office that Thursday at 10:30AM and we'd talk about it. When I got there he said I'd start with him at the beginning of summer, working on a problem involving electron scattering by copper atoms in silicon, along with a post doc, Frank Blatt. I was completely surprised and had no intention of doing theory at that point, but it sounded intriguing and I figured it wouldn't hurt to give it a try, even if only for the summer. I never looked back. We got together with Frank to talk about this numerical calculation, in those days with a hand calculator. They were discussing what silicon effective mass to use in the calculation, and I said it seemed to me we shouldn't use an effective mass for this, but the real electron mass. After some discussion they decided I was right, and I think it helped me feel that I belonged there.

I signed up for a student-staff apartment right next to the campus for Lucky and me, starting late summer. My father decided to give us a new car for a wedding present and I bought a beautiful blue and white four-door Bel Air Chevy. I loved the old Chevy but, just as when I started doing theory rather than experiment, when I left the old Chevy in the parking lot to take the new car I never looked back. In June, I headed back to Toledo with the new Bel Air, and Lucky and her friend Nicky started their journey from California in her old Dodge. It was often boiling over when they were coming across the desert but they almost made it back to Garden City. Then the car finally died in New Jersey, close enough that Lucky's father, Al, could drive out and pick them up, and bring their things over.

I drove east for the wedding, as did my parents, and as did Chuck and his new friend Carolyn Warner. We had a festive rehearsal dinner

Our families at the wedding, my father and mother, Charlie and Gertrude, Lucky and Walt, and Lucky's parents, Marion and Al Carley.

Lucky throwing the bouquet. The two ladies at the right are Lucky's roommate, Nicky and my brother Chuck's fiancée, Carolyn Warner.

Lucky and Walt leaving the wedding reception.

and Lucky had a last-minute shower the morning of the wedding. The wedding went well followed by a reception in the Carleys' back yard and Carolyn made sure that she caught the bridal bouquet. Indeed they were married the next spring.

There was one complication about the wedding of which I wasn't aware. Lucky's grandmother, Marian's mother, lived with her sons Harlan and Willis Prince in Brooklyn. Harlan was a little weird and when I first met him he asked me what I did. I said I was a physicist, and he said: "Oh, like Einstein, that dirty communist rat." I said something like: "Oh, come on! He's no more communist than I am." That wasn't a good idea, because Harlan took it as a confession and sent a letter to that effect to J. Edgar Hoover, the FBI. That turned up the next time when my Q-clearance for the Atomic Energy Commission needed renewal, probably for Sandia Laboratories. I was *thoroughly* investigated and an agent even came to talk to Lucky in person, an unusual procedure, but I passed. Harlan didn't come to our wedding, but Marian was very nervous that he might show up for the "Does anyone present know of any reason why this man and this woman..."

After the reception Lucky and I headed off to Connecticut where I had a motel room reserved. When we got there in the evening, the manager came out to give me back my check; another customer had come along and would stay a week. I subsequently tried everything I could to get AAA to remove his recommendation, but it didn't help us that night. Everything was sold out locally so we had dinner and drove back to Stamford, Connecticut, where we could get a hotel. From there, we went north to Quebec City. My father had enthused about eating at the Frontenac Hotel and we did. We came down through Montreal, we visited Niagara Falls, and then on to Sages Grove where we stayed in the streetcar. On the weekend my family came and we were off to Illinois.

Lucky had been offered a first-grade teaching job in Champaign, a short drive from our apartment. I got a desk in a big office in the Physics Department with a group of other theory students. It was called *The Institute For Retarded Studies* and was inserted between the second and third floors. It was a mixed group, Sol Gartenhaus, Dick Curtis, and Jerry Franklin in particle physics, Bob Schrieffer, Danny Mattis, and me in solid state. I was finally a part of the Department. We were very happy with our first-floor apartment in the Student-Staff complex. They even had a workshop in the basement and I made some bedside tables for our apartment, my first in a series of furniture making.

We also had friends there whom Lucky had met during teaching. One set, Barbara and Ralph Albrecht ended up in San Carlos, California, and we've remained friends. Another set, Joyce and Ted Puckorius, were living in a trailer in a trailer park. We were there for dinner one night when he kept filling our martini classes until we were both quite drunk. As I remember, we were both outside upchucking, and had to lie down a while on their bed before returning home. We've never again repeated such a scene.

Chuck had gotten married a year after us, so we were back in Toledo for the wedding. My father was quite sick at the time, but attended the wedding and died not so long after. Lucky's parents left Garden City and moved to Vermont, buying a beautiful old house with gingerbread trim on a farm outside of Middlebury. We went back east to visit them there.

Back in Illinois, I had completed my part of the scattering-time project, and another small topic on radiation damage. I was proud of my contribution on radiation damage and Seitz suggested that we do joint papers at the March American Physical Society meeting. We *did* have joint ten-minute papers scheduled, Seitz and Harrison, and then Harrison and Seitz, but it turned out Seitz just went seamlessly from one to the other, talking for twenty minutes. In any case, I was ready to pick a research topic. One possibility Seitz suggested was very appealing to me, figuring out just what kind of a metal carbon would make if you forced the atoms to be in a metallic crystal structure, but I was afraid the results might not be interesting to anyone else. Another possibility was electron scattering by high-frequency (called optical) vibrations in semiconductors such as silicon and gallium arsenide. The transistor had only recently been invented and semiconductors were becoming a hot topic. That was more in the mainstream and I opted for that. It was a good choice; the study and use of semiconductors was about to grow into a huge field and we were at the beginning.

The first order of business was to understand optical lattice vibrations. I approached it as it turned out I would always start: dig in as if no one else had ever worked in the field. I knew enough from the solid-state-physics course to know how to start, and I imagined the atoms connected to each other with little springs. I could tell how stiff the springs should be by calculating the stiffness of the crystals, which had been measured. Then I could calculate all the vibration frequencies for silicon and other semiconductors. That was a big project, and when I was done I learned that many other people had done similar calculations, but I never regretted doing it myself rather than reading other people's papers on it.

A principal figure in this earlier work was Max Born, a German physicist in England who, with students, had calculated the vibrations in diamond, very similar to silicon. My approach was almost the same as his and the results were very similar. At the same time, an East Indian physicist, Sir C. V. Raman, had done, with students, the same calculation with far different results. Raman had been knighted for discovering an important experimental effect (the Raman Effect) involving interaction of light with these vibrations and this led him to the conclusion that he was omniscient in these matters, but he was dead wrong. However, I think he was the only Nobel Prize winner in India at the time and had enough control over his field that one could succeed in theoretical physics in India only by writing supporting papers. I could see immediately why almost all of these papers were completely wrong. But there was one man, named P. Krishnamurti, who had managed to write a paper with nothing being wrong yet still leave Raman thinking it was in support of him. I've never met P. Krishnamurti, but I have heard that he became a highly respected physicist in India and I'm not surprised.

I even found a flaw in Max Born's approach. There is one kind of spring between atoms, between nearest neighbors that keeps them a particular distance apart. There is another which keeps the angle between two neighboring atoms constant, but that one includes the relative position of two atoms which are *second* neighbors to each other. Born included only the nearest neighbors and the resulting forces were such that if the entire crystal were tipped and released, it would oscillate back and forth around the starting orientation. I corrected it in my approach and published the results and Born's error did not matter that much. My approach became known as the Keating model after another graduate student who had read my paper and used it in his thesis. He once apologized to me for that, but it wasn't his fault: he properly referred to me but those who followed him didn't look that far. I don't know how easy it would have been to have gotten all of this straight if I had started reading Born and Raman at the beginning rather than working it out first by myself. In any case, I had worked it all out on my own and that is the way I began new areas for the rest of my life.

In his study, Born had used group theory, not so well known then and not taught at the University of Illinois. The main text was by Eugene Wigner and in German. Actually Wigner had directed Fred Seitz for his thesis, making him my "academic grandfather". Our group in the Institute for Retarded Study decided to teach ourselves, each translating some of Wigner's *Gruppentheorie* and giving a talk for the others. The book was very mathematical and formal, and not to my taste, so I instead found a book on applications of group theory to solid-state and molecular problems, by Bhagavantam, another Indian, but without so many mistakes in it as with Raman's people. That *was* to my taste and I included those applications in my thesis and later at the beginning of my solid-state text. For me, it appeared to be the only useful part of our joint effort to learn group theory.

I'm not sure how often I went to the weekly Physics Colloquium, but one that I remember quite clearly was given by Brian Pippard, visiting the University of Chicago for an extended period from Cambridge University in England. He was doing some very careful experiments to determine the Fermi surface for copper metal. People had been trying for some fifty years to understand the electronic structure of metals, making up models and fitting them suitably for experiment but nothing that could be called progress because the model for each experiment was different. Pippard had recognized that the Fermi surface was something that really existed and could be measured and that was central to the electronic structure. I'm not sure if I recognized the importance of that step then or only later on, but it became central to

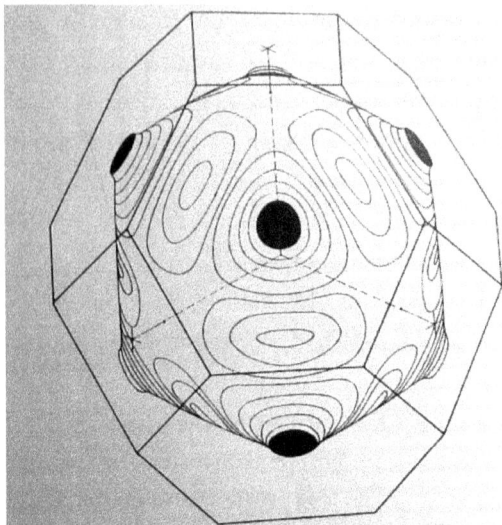

The Fermi surface of copper,
according to Pippard.

my thinking in a few years. Much later it turned out that there was another shape, similar but more bulging that would also fit Pippard's data and that turned out to be the right one. More importantly, the big step was recognizing that the Fermi surface is central and can be determined.

I met some interesting figures in physics through Seitz. One was a Professor Welker, visiting from Germany, who had invented gallium arsenide, one of the compounds I had been studying for my thesis. Another was Bill Shockley who joined us for breakfast at an American Physical Society meeting which I attended in my last year at Illinois. That was the year that Shockley won the Nobel Prize with John Bardeen, whom I knew at Illinois, and Walter Brattain from Bell Telephone Laboratories. Shockley ended up at Stanford and was a friend till his death.

I had noticed that every time someone in our office came to graduate, they were terribly stressed when writing up their thesis. In those days it was also necessary to hire a professional to type up the final thesis. I therefore kept writing up my work as it progressed, a practice which I continued from then on. Finally in the midwinter of my third year, Seitz said he thought it was time to complete it for graduation that June. He was quite surprised when I delivered the entire thesis to him in a week or so. He read it over carefully and marked it up extensively. He said a proper thesis should read like the New York Times, so we converted it to that style. I thought it read much better before, but of course, I went along. He also then suggested some extensions of the work so I couldn't avoid the rush as I first intended.

During my last year I began interviewing for jobs. I enjoyed the chance to travel but the first interview, at Lincoln Laboratories in Massachusetts, was a little unsettling. I was met by a man who said he was delighted to finally meet someone who had done some good experiments on rare-earth alloys. I said, "Well, that's not me. I do theory." He paused and said, "OK, anyhow let's go get a cup of coffee." I *was* shown a piece of gallium arsenide on that visit, which I had never seen, but otherwise the place appeared quite dull. The RCA laboratory in New Jersey was very different. Frank Herman showed me around and was anxious for me to come to RCA. In the end, he was never able to get RCA to offer me a position and has always felt badly about it. I thought I did have an offer from RCA and felt badly that I decided against it. We've been good friends ever since, partly based on this mutual guilt.

Bell Laboratories, Westinghouse, and General Electric did each offer me a position. It seemed clear that Bell was the most distinguished lab and the best career choice, but I liked the people at GE so much and felt so comfortable in the Metallurgy and Ceramics Department that I accepted their offer. That was another decision I never regretted.

When June came, I arranged the traditional graduating party in the party room of our apartment building, but not with the traditional pineapple punch with rum. I got a keg of beer, as in Cornell days. I was quite pleased that John Bardeen came, and also noticed that he never strayed far from the keg. I don't remember if Seitz came.

For moving, we had arranged for the packers and the van to come, since General Electric was picking up the tab for our move east we took everything. We even took the brick and board bookshelf that we had built in our living room. The packer put the bricks in a box on the middle of the floor and never moved it. When the movers came, each one at some point would start to pick it up, but then move on to something else. It was the last item to go. They packed everything, even the ashtrays with cigarette butts in them. When they were gone, Lucky and I headed west for a one-month trip on the way to Schenectady. We looked up friends on the way, in Los Alamos and Spring Valley in California, in the Bay Area and, on the way back, in Chicago and of course Toledo. We arrived in Schenectady in time to start at the lab on July 3, leaving the July 4 holiday available to settle in.

Chapter IV

GE IN SCHENECTADY

We rented a one-bedroom apartment in a complex on Daisy Lane in Schenectady, a couple of miles from the General Electric Research Laboratory, which is on the banks of the Mohawk River. It was a fine place to be and we soon had friends among our neighbors on Daisy Lane, and even another member of my Section at GE, Bill Johnston, to carpool with. Our furniture, including the board and brick bookcase, was soon in place and we settled in. Lucky was acquainted with a Norwegian girl, Inger Giaever, who had just arrived with her husband Ivar and a young son, John. They were to increasingly become an important part of our lives.

Lucky also met a local girl, Rosella Suits, and we played bridge with her and her husband Jim, sometimes until very late at night. We continued to be in touch with the Suits and they later moved to California where we renewed our friendship with their family. Jim's father, Guy, was the Director of the General Electric Research Laboratory and an old friend of Fred Seitz. He was interesting. He owned a summer house (called a camp in upstate New York) on Jenny Lake and considered himself the mayor of Jenny Lake. I think the other residents treated him as one. But when he bought an airplane, with pontoons so he could visit the lake by plane, the residents got together and said no to landing on the lake. Guy abruptly sold the house on Jenny Lake and bought one on

Lake George, where he was free to land when he liked. We stayed at the Lake George house one or two times.

The head of my Section at the lab was John Fisher, theoretically inclined, but with a metallurgical background. When I asked him some time along just what I was supposed to be doing there, he responded: "The best science you can do." I certainly didn't feel restricted. That was consistent with what we had heard from a Swiss member of the Section, Werner Känzig, a friend from Illinois who had moved there the year before. He said: "GE is wonderful! They let you come into the lab and work all night." The section was called Physical Metallurgy, but was really a physics section. I was the new young theorist and had an early chance to make an impression. We were at Werner's house one evening and looking at a bronze horn that was on the coffee table. Someone wondered just how it was made and I looked and said: "I think it's a slush casting." I'm sure it was, because we had learned about them in a metallurgy class at Cornell. They pour the molten bronze into a steel mold, slosh it around, and pour out the part that didn't freeze to the mold. Open the mold and out comes a hollow bronze with the outside shaped just like the mold. However coming from the young theorist it astonished the physicists who knew little about metallurgy.

Usually, most young PhD's after graduation continue with their thesis topic, finishing up some things they hadn't done before. I didn't want to do that, so I switched from semiconductors to the theory of metals. I had a way of going at it by dividing the metal into cells, each with one atom, and matching quantum-mechanical wave functions from one cell to the next; it was a little reminiscent of an earlier approach by Seitz. It worked well enough that I published a couple of papers the next year, but then I decided it wasn't *that* good and went on to other things. By then I had quite a stack of calculations, maybe an inch or two thick, and I picked them up and dropped them in the wastebasket. It seemed therapeutic at the time, and I think maybe it was.

Every day at lunchtime a number of us on our corridor gathered together and walked to the cafeteria for lunch, in summer to a table out behind the cafeteria. That was great from my point of view. It was a

Ivar Giaever explaining his work to Walt, Charlie Bean and John Fisher at the General Electric Research Laboratory.

collection of interesting people with interesting views and we could settle the world's problems, among other things. John Fisher in particular had a novel and innovative view on most topics. I remember one time he asked everyone to close their eyes, and then try to remember which of us was wearing glasses. It turned out that we all did very poorly.

Another friend was Henry Ehrenreich, whom I had known somewhat at Cornell, a physics major a year ahead of me. He had gotten his PhD and come to the General Physics Department in GE, and bought a house in a suburb, Scotia, across the Mohawk River. We were invited to dinner there and found the house and the neighborhood very appealing. We went to talk to the builder, Walter Sosha, who had other lots in the neighborhood. We ended up contracting him to build a house for us down the street from Henry, who liked to say that our contract read "To build a house just like Henry's, with the following improvements...". It was mostly true; once during construction, we noticed that the hearth didn't continue to the front wall as Henry's did. We spoke to Sosha about it and he agreed, if Henry had it he would build it into ours. The principal carpenter for Sosha was Chet, who drank beer continually during construction, frequently leaving the quart empties between the studs. I'm sure a number of them remain there even today. Years later I was flying to a meeting with Henry and he said my being there made him feel safe. Everything happened in my life one year after his, so if we were in a plane together, it couldn't crash. He *was* married a year or so before me and *did* leave GE for a university (Harvard) a couple of years before I did, but he died some years ago, so the rule didn't save him.

Back at Daisy Lane, by Christmas time we had enough friends to host a hot toddy, or hot buttered rum, party one weekend. Lucky was pregnant with Rick and he arrived on April 6. We got a baby carriage where he slept till we moved into the house. Our neighbors in the next apartment, the Semmels, had a baby earlier that year and kept their apartment unpleasantly warm after. We decided that that was inappropriate and we opened the window as usual the first night Rick was there. It scared us to wake up in a cold room, but Rick was fine.

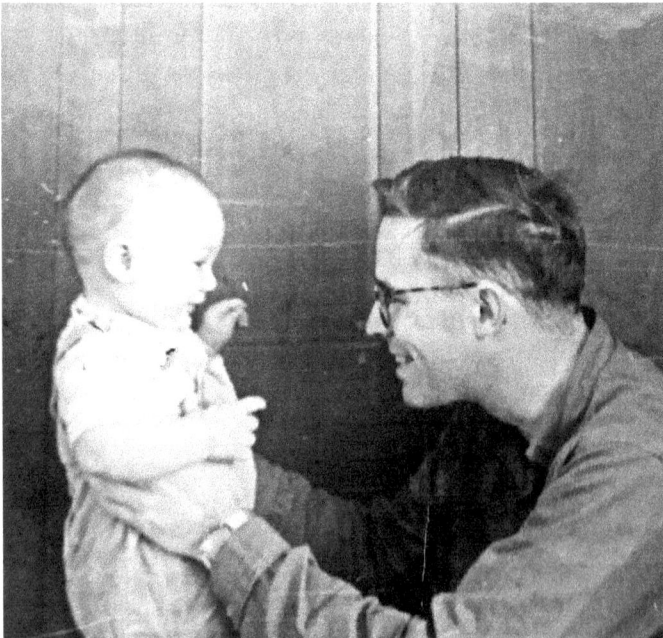

Walt with Rick in 1957.

Lucky and Walt with brother Chuck and his wife Carolyn on the right, in Hackney's Restaurant in Atlantic City.

My mother was anxious for Chuck and I to stay in touch and planned a vacation for all of us in Ocean City, New Jersey. Both Rick and their Charlie were babies but we were right on the beach, so we enjoyed ourselves. One night my mother babysat and the four of us went off to dinner on the boardwalk in Atlantic City. By then, my father had died and we no longer had the streetcar, but for another year my mother rented a house in Chaska Beach on Lake Erie close to Sages Grove and we all met there. Then Carolyn's father bought a house at Chaska and we all went there in the summer for a number of years. My mother generally brought her cleaning lady, Flossie, and she also served as babysitter. Ginger Luscombe, who was part of our earlier group at Sages, had since married and also rented a house at Chaska with her two children. It was like the old days at Toonerville.

I had a chance during these visits to get reacquainted with Chuck. He was a little competitive and always wanted to know what my salary was. I think he stopped asking before he fell behind. He had started out as a successful engineer for Surface Combustion Company in Toledo, but shifted from job to job after that, including for a while as manager of a local television station. He also drifted more and more to the right politically. I remember him at Chaska Beach chiding us Californians for having a movie star as governor. I told him to wait, at sometime Ronald

Reagan would be running for president and Chuck would vote for him, but I wouldn't. That was one of the few times my prediction of the future turned out to be spot on.

We moved into our house at 40 Cedar Lane, Scotia, at the beginning of July, in 1957. It was a split-level house, comprised of six levels, starting with the basement. The next level had the laundry, a playroom which we later finished, and a garage. The third level, a few feet above ground level, had the living room, dining room, and kitchen. The fourth level had three bedrooms and a bathroom. The fifth had a large bedroom with a bath which we later finished, and the sixth was a small attic, all of this for some $21,000. We had to install a lawn, which in the back melded into a small forest before the open fields of the farm of the county old-folk's home. Several houses down, the farm dropped off as a nice sledding hill; and in the other direction, it led to a stream with a small waterfall. It was as good as it sounds. In two years, John was born to make us four.

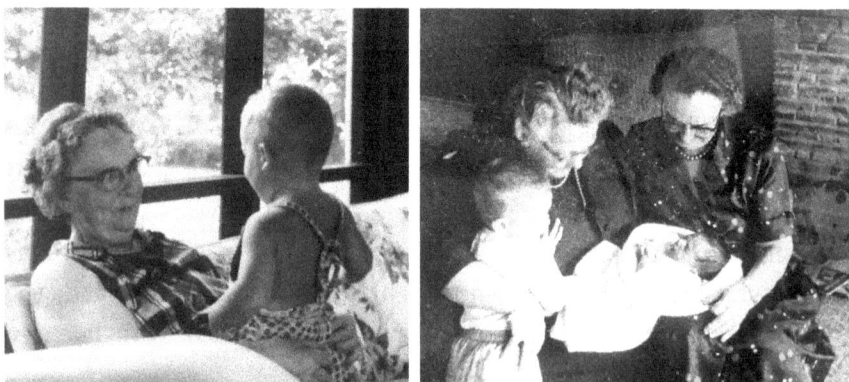

My mother, Gertrude, with John, on the left. Lucky's mother Marian and Gertrude with John and Rick on the right.

In the following years we had lots to do with our new house. I had to build a work bench in the basement, and that area was later also used when I started making beer. It wasn't very good beer, but it sure was cheap. These were before the days when making craft home brew became popular. I remember when it did, reading about it in the paper.

The article said how fine it was and how different from the old days when one poured a can of hop-flavored malt extract into a crock with five gallons of water and five pounds of sugar and added Fleischmann's yeast, exactly my recipe. I waited till the mixture stopped bubbling, meaning the sugar was all converted. Then I added a little more sugar, filled and capped the bottles. The extra sugar was to allow enough further fermentation to carbonate the beer. I remember one night, while we were listening to John F. Kennedy talking about the missile crisis in Cuba, and explosions started in the basement which added to the scare; I'd added a little too much extra sugar. I rushed the cases out to the back yard and the explosions subsided.

We did finish the playroom next to the garage, one year after Henry Ehrenreich did his. Henry bought random vinyl tiles of different colors of cork, and hired someone to install them. He wanted a random pattern so before the workman came he drew up a diagram of the floor and used a random-number table to place the different-colored tiles. It looked terrible, big blotches of dark and light. He didn't know what to do, but the workman who came the next day said it was no problem. He had a random four-by-four pattern that he put down, each set rotated compared to the last set. It looked fine. I put the tiles down myself, in a way that looked random to me, and it also looked fine.

By this time, at the lab, I was still thinking about metals though with a different approach. This expanded beyond the lab; in the summer there were week-long Gordon Conferences on different topics at prep schools in New Hampshire, and I had gone to one on metal physics. One high spot was when a very famous chemist, Lars Onsager, showed up before dinner with a bottle of gin. For some reason I had a bottle of dry vermouth and we became buddies for the conference.

When a colleague at GE, Ben Roberts, told me he was going to stop the experiments he was doing and wanted to try something new, I remembered Pippard's colloquium at Illinois, and suggested he might try to determine Fermi surfaces of metals. I had also talked with a friend at a Gordon Conference, Harvey Brooks, Dean of Engineering and Applied Science at Harvard, who thought it was time to pick a metal and try to understand everything about it, as silicon had been picked for semiconductors, and he chose aluminum as the metal. So I suggested aluminum to Ben, and I had read about measurements of ultrasound attenuation in metals in a magnetic field that had the prospect of measuring Fermi-surface diameters, so I suggested he take up such measurements.

Aluminum would be expected to have a complicated Fermi surface with several pieces, rather than the sort-of-spherical surface of Pippard's copper, that I showed in Chapter III. This was because aluminum had three available electrons per atom whereas copper had only one. Actually, a student at Cambridge University, named Andrew Gold, had been studying the Fermi surface of lead, with *four* electrons per atom, by a different method. He was helped in interpreting his results by starting with a spherical Fermi surface, like copper but four times the volume for the four electrons per atom. Then this big sphere, which

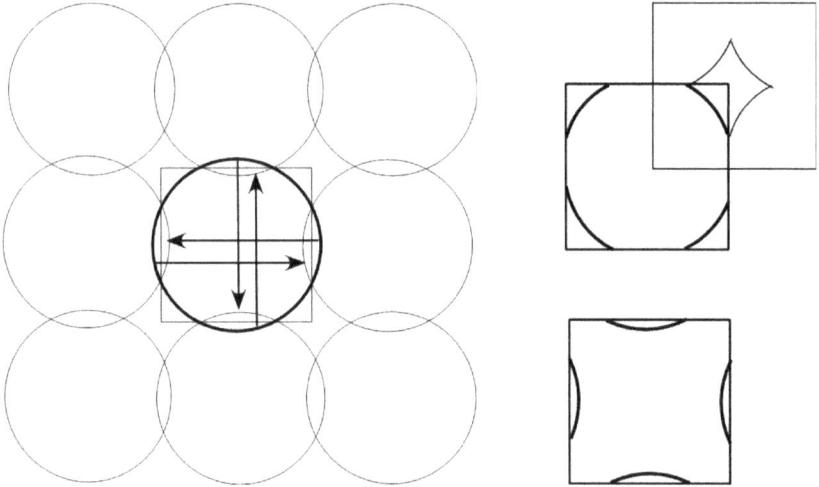

Construction of Fermi surfaces. The heavy circle at the center on the left represents a free-electron Fermi surface in wavenumber space. In solids it is useful to limit the range of the wavenumbers to a Brillouin Zone, represented by the square. Then the pieces outside the box are translated into the box, indicated by the arrows with lengths equal to the dimensions of the box. The resulting Fermi surfaces are shown to the right. The simpler way is simply to reproduce the sphere on a lattice, as shown with light lines on the left. It can also be helpful to shift the Brillouin Zone to another position, putting the center of the square at what was the corner, as seen to the upper right, to bring the pieces of the Fermi surface together.

corresponded to a free-electron gas, was broken up into pieces by the presence of the crystal lattice. It was an intricate procedure, but he found a piece which seemed to fit his data, though it turned out that he chose the wrong piece. Another physicist at Cambridge, Volker Heine had thought about aluminum in the same way. This looked like the way to proceed.

The procedure for getting these Fermi surfaces was intricate because it involved moving parts of the sphere toward the center of the sphere in an appropriate pattern. It occurred to me that this was unnecessary; one could simply construct a number of spheres around appropriate points and see how they intersected and overlapped in the center. This is illustrated in the figure. This made the process so simple that many people could understand it, and it came to be known as the *Harrison Construction*. I realized that you don't get famous by doing something complicated.

This was quite straightforward so I did the construction for three electrons per atom like aluminum, and four like lead, and two like calcium. Those three metals have the same crystal structure. GE sometimes brought students into the lab for summer jobs and I got such a student, Frank W. Warner, III, to do the corresponding constructions for

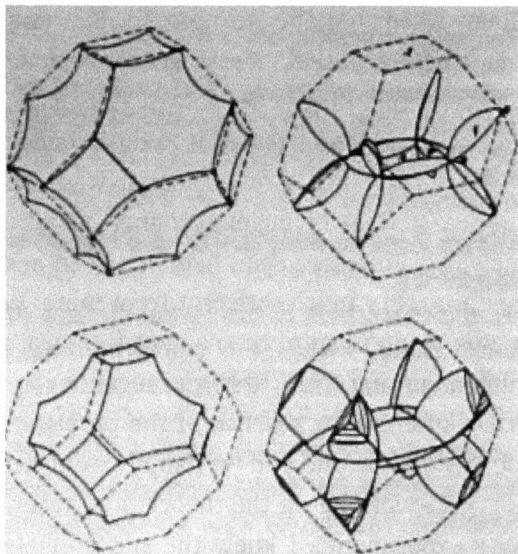

Main free-electron Fermi surfaces for aluminum, above, and lead below drawn by F. W. Warner III.

two other common crystal structures of metals (body-centered-cubic and hexagonal-close-packed) for one, two, three, and four electrons per atom.

Frank was overly precise, but maybe that's OK; nobody redoes the constructions, but they do utilize his figures. I gave him a quick perspective to use without realizing that it distorts figures a little. One day I came in and he was working on a surface for one electron per atom which is exactly spherical (as in the cover picture here for gold), but not in my perspective, and he was constructing the slightly oval circle to represent it. Also for better precision he constructed each figure on tracing paper larger than 8½ by 11 inches. When I assembled the four-by-four sets it was necessary to send the composite down to the Large Steam Turbine drawing shop in the GE main plant to have it reduced.

Ben Roberts was getting some successful measurements of the aluminum Fermi surface, and decided it would be cool to have such a surface (upper left in the figure) machined from aluminum. He placed an order with the machine shop and a month later discovered they had charged some thousand dollars to his account. They had given it to an apprentice who had done well, but found that after a few cuts for producing concave surfaces there was no way to hold the piece that remained. He had to machine a convex surface on another piece and cement them together to proceed. Ben was able to spread the charge over a few months so it wasn't conspicuous and it *was* a beautiful surface, about four inches across. They gave it to me when I left GE, but more recently I contributed it to the Schenectady Historical Museum.

At some point a new theorist was added to our section, Charlie Duke, a student of Eugene Wigner, which made him my professional uncle; my Professor Seitz had also been a student of Wigner. Charlie's style was very much different from mine. Whatever question came up, he was ready to go to the computer to deal with it. Mine was always to think about it a while first, and then maybe try to make a simple model that contained the essence of the problem. However, he seemed to be good at what he did and we happily went our own ways.

I was working on the theory of ultrasonic absorption in support of Ben's experiments. I knew the states of electrons in a magnetic field from quantum mechanics and was making calculations of their absorption of ultrasound waves. I was having some success, and gave a couple of talks on the work, when a consultant, Morrel Cohen from the University of Chicago, visited. He asked what I was doing and I told him. He said that was interesting; he had just assigned a graduate student, Mike Harrison, the same problem for his thesis. I asked how they were going about it and he explained a purely classical approach, as opposed to my quantum approach. That was totally new to me; I really hadn't learned classical physics, but he explained his path-integral method. After he returned to Chicago, I continued applying this method to all the cases I had treated, and it worked beautifully. I had learned my lesson and put all this away. It was Mike's problem and Morrel's method. When Morrel next visited, I told him what a fool I had been, doing it with complicated quantum mechanics when the classical theory worked so well. He said: "Did it really? Mike is just getting tooled up to start." I showed him all of my results and he said: "It seems silly for Mike to redo all of those calculations. Let's write it up and publish it jointly." We did, M. H. Cohen, M. J. Harrison, and W. A. Harrison, alphabetical as Morrel always preferred. It was a good paper and when it led to an invited talk at an APS Meeting, Morrel gave it. Morrel had the insight and the method, I did all the work, and Mike had squatter's rights. I expect some readers thought I had a wife, Mary Jane, when they saw M. J. Harrison as a coauthor.

John Fisher wanted to have some experiments on current through thin oxide films, which he thought would be from quantum-mechanical electron tunneling. He hired Ivar, our Norwegian friend from Daisy Lane, who then got a lab in the bay next to my office to do the experiments. We became good friends and he came to my office every morning for coffee and discussions. He also began taking graduate physics courses at Rensselaer Polytechnic Institute, RPI, in neighboring Troy, New York. He recognized that one didn't get respect at the lab without a PhD. The experiments weren't so impressive. They were consistent with the assumption that the electrons were tunneling, but equally consistent with the assumption that there were pin-hole leaks through the oxide, which some people believed was the case. Ivar took a solid-state course from Hilbert Huntington and heard that there was a gap of missing states in a superconducting metal. He thought maybe he could detect it with his experiments. The gap idea was from then the very recent theory of superconductors by Bardeen, a post doc at Illinois named Leon Cooper, and Bob Schrieffer from our Institute for Retarded Study, the BCS theory. Ivar asked me, and others, whether he would be able to see it. I thought not; the argument for seeing it remained true no matter how thin the oxide, even zero thickness, and we knew there was no effect with no oxide. However, I urged that he try it in any case.

When he did the experiment, the result was spectacular. The gap was clearly visible and one of the very first experiments to see it directly. We later learned that as the oxide gets thinner a *new* mechanism allows the current to flow, the Josephson effect, so I had been wrong in a complicated way. In the end Ivar, Brian Josephson, and Leo Esaki shared a Nobel Prize on tunneling. I later spent two weeks one summer in Esaki's group at IBM and Brian Josephson became a long-term friend.

I worked out a theory of tunneling in systems such as Ivar prepared, but had trouble on one aspect of the calculation. When Bardeen came as a consultant I asked him what he thought, but he just mumbled something about doing it in a different way and it didn't help me. Some weeks later Bardeen sent me a short manuscript, "Tunneling from a Many-Particle Point of View". He felt uncomfortable about it because it was my problem. I assured him that his treatment was much different from anything I might do, and to go ahead and publish. I soon published "Tunneling from an Independent-Particle Point of View". It was applicable to tunneling systems in which there were no superconducting metals and I think mine proved more useful in the long run.

There was an interesting sidelight; I think all this led Bardeen to make the only physics mistake I know from him. On the day he received page proofs for his paper he also received a preprint from Brian Josephson on his new effect. Bardeen glanced at it, decided it was wrong, and said so in a "Note Added in Proof" to his paper. This remained uncertain for a time, and at a physics meeting in London, which I did not attend, Brian gave a talk on his effect. Bardeen was there and needed to comment on the subject. The audience was very impressed at Brian's calm certainty as he argued in what they could recognize as a somewhat lower-class accent with this double Nobel Prize winner. Bardeen later conceded that Brian was right.

I enjoyed going to one or two Physical Society meetings a year, and seeing old friends from Cornell and Illinois, as well as GE friends. On one occasion in the New Yorker Hotel bar I ran into Arthur Kantrowitz, from whom I had taken thermodynamics at Cornell, a course I greatly enjoyed. He taught that the first law of thermodynamics was that you can't make a perpetual-motion machine of the first kind. You can guess what the second and third laws were. My father really only knew the first law, which was basically conservation of energy, but if someone thought they had invented a perpetual-motion machine he would always think that he could tell them why it wouldn't work, that it violated the first law. Mr. Archambo's bathmat dripping water violated the first law.

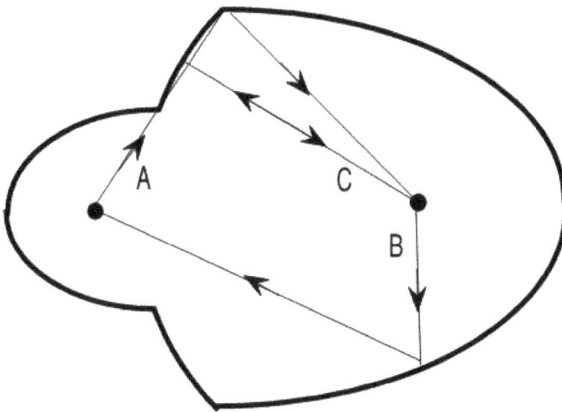

The Chinese flatiron perpetual motion machine. Two particles are the focii for a small partial ellipsoid reflector on the left and a large partial ellipsoid on the right. The spherical mirror between them is centered on the focus at the right. In thermal equilibrium both particles emit radiation at the same rate. All of the radiation from that on the left, beam A, strikes an ellipsoidal surface and reflects to the other focus. (It's a mathematical property of ellipsoids that radiation from one focus reflects to the other.) Some of the radiation from the right, beam B, reflects to the particle on the left, but some, beam C, strikes the spherical surface and reflects back to the particle on right, which brings it to a higher temperature (contrary to the second law of thermodynamics). The temperature difference can be used to generate energy, for example with a thermocouple.

The second law was more subtle, roughly speaking, that heat always flows from places of higher temperature to places of lower temperature. I learned of a perpetual-motion machine of the second kind called a Chinese flatiron from, I think, a math professor at Cornell. There are two elliptical mirrors, with the same two focal points, joined by a spherical mirror centered on one focus, as shown in the figure. If a small object is placed at each focus, both will radiate thermal energy, and ordinarily will absorb thermal energy at the same rate if everything is at the same temperature. However, elliptical mirrors have the property that radiation from one focus will be reflected entirely to the other focus. Thus all the radiation from the left object will arrive at the right object, but some 20% of that from the right object will strike the spherical mirror and be reflected back to the right object; it receives 20% more radiation and will heat up while that on the left will receive 20% less and cool down. That violates the second law because the temperature difference could be used to generate useful energy, a perpetual-motion machine of the second kind. It occurred to me that someone like my father would be puzzled because he thinks only in terms of the first law and indeed all the heat energy added to the second object was removed from the first. He might be puzzled and ask: "Have you tried it?" The clever answer would be: "Yes, but it didn't work." He would be relieved, but then you add: "I connected it to a light bulb, and the light-bulb lit up at first but the mirrors got fogged up; I think I can fix it." Then thinking again of the first law he might say: "Of course, if you take energy out to light up the bulb, the apparatus must cool down." Having contributed to the theory of the device he might believe that it works. I actually tried this out on a Turkish post doc at Illinois, and at the end he was ready to try to make one. I'll say at the end of the chapter what is wrong with the argument, but don't want to spoil it for now. Back to Arthur Kantrowitz. While we were sitting at the bar talking, smoke arose from behind the bar. The barmaid shrieked. Arthur reached over the bar, picked up the seltzer-water bottle and put the fire out. He was a practical man.

In 1964, Ivar and I decided that it was time for GE to give us a European trip. We got a list of conferences and found one in the summer in Copenhagen on lattice dynamics, the subject of much of my doctoral thesis. GE okayed it and we planned our first trip to Europe. Lucky and I had a lady who could take care of the boys while we were gone and for some reason we flew out of Montreal. The first stop was London, with a side trip to Cambridge. Then across the continent, ending up finally in Copenhagen, where we met the Giaevers who had been visiting back in Norway. Lucky and I were there first and went to dinner the first night by ourselves. We ordered aquavit as we had learned from the Giaevers, and the friendly waiter left the bottle at the table. We didn't want more, but maybe had a second. In the end they charged us for a whole bottle. We told the Gaievers about it and Ivar laughed; he said they do that to Americans, but they would never do it to a Norwegian. The next night we four had dinner and the same thing happened! Another night we wanted to go see Nyhavn, the seedy part of Copenhagen, and Ivar was adamant against it. Lucky and I went anyway and had a beer, and nothing happened.

It turned out to be an extraordinarily interesting conference for me. Every well-known person in the field was there. I even met Max Born whose work I discussed here earlier in connection with my thesis. There was a reception where we met him on the way in. For some reason I didn't mention my thesis, but a book *Mr. Tompkins in Wonderland*, a fanciful book on modern physics and quantum mechanics I had read in high school. He remarked: "Oh, by George Gamov, that scoundrel," or something stronger. I was surprised but he explained how, much earlier, Gamov in the Soviet Union was invited to an important conference in Europe and was not allowed out by the communist regime. Gamov appealed the decision and asked for support from distinguished Europeans such as Born. Born vouched for him, said that it was important for him to attend, that he would learn about western physics, could contribute to the subject, and return to Russia. Gamov *was* allowed to come, but he never went back. The old German Born felt that it was a black mark on his character, having vouched for him and then having him not follow through as expected. My talk went well, but included an alternative explanation of some features ordinarily attributed to the Kohn effect. Walter Kohn was there which made for some lively discussion, but we were invited, with a number of others, to Kohn's temporary quarters in Copenhagen one night for drinks.

Back home John Fisher developed a new interest. He was excited by the new BCS theory of superconductivity and decided that something very similar was behind all of the fundamental particles that make up the world. That wasn't crazy but I wasn't in favor of it. I could tell that even if he was right and developed such a theory, none of the workers in that field would listen or notice it. He continued the program to the day he died in 2018, and no one was listening.

And in fact Peter Higgs and François Englert won the Nobel Prize for such a theory in 2013. I was very much interested in that prize. In 1961, I went to an APS meeting in Monterey. One evening at the meeting I went out to dinner with Bob Braut and the student François Englert. As I remember, Braut mentioned that he had heard John Fisher talk on the subject at the meeting and found it interesting, which surprised me. During dinner, François excused himself to go to the men's room and that was the last I ever saw of him. Some years later in East Germany I was at a meeting where there were name cards at each place and I noticed one lady had a card with Englert on it. I asked her if it was possible that she was connected with François, and she said he was her brother-in-law, that he was well and living in Belgium. Just at the time we were in Monterey, Braut and Englert, and I think Higgs independently, were developing the theory for which they got the prize. By then Bob Braut had died so he did not share it.

Roland Schmitt replaced John Fisher as head of our group. He decided that GE should sponsor a physics meeting, and with all this activity why not on Fermi Surfaces. So in the summer of 1960 we held the Fermi Surface Conference in Cooperstown, New York, with all the principal workers attending and Barney (M. B.) Webb and I doing the proceedings. Ben Roberts got interested and decided to make plastic duplicates of his aluminum Fermi-surface model. It turned out to be quite a chore because the epoxy plastic he used heated up when it hardened and with a piece that large it was slow to cool. However, he managed to produce the needed 120 or so and they were prized by all of the participants.

The conference wasn't arranged as I would have chosen. The principal speaker on each topic was an older leader, while I would have chosen the young workers who were doing the research. However, it worked beautifully, with each speaker nicely summarizing all of the work going on in his area. My Fermi surfaces for the large array of metals played a significant role in the conference. It was the most exciting conference that I had attended, or ever did attend. Everything was new and happening fast. A few years later I chaired a Gordon Conference, and did it my way with young workers being the principal speakers. It was very successful, but nothing was new nor happening fast and the result was not really memorable.

GE sent a recording crew to cover the Fermi surface conference, with microphones all around the room. When we came back to produce the proceedings, I had the tape and a foot control and typed up all of the discussions following the talks; the speakers turned in manuscripts of their talks. I'd listen carefully to each comment, leave out the "uh"s and "I mean"s and put down what the speaker intended to say. One speaker, John Ziman from Cambridge, was visiting GE after the conference and I gave him a copy of his comments. He said: "Oh, no, that's not what I said at all!" I had him come and listen to his bumbling comments as recorded, and he said, "Oh Pshaw! Go ahead and write what you want." I even took the figure of Pippard's Fermi surface from another talk and removed a distortion he had unintentionally made in the original drawing to obtain the version shown here in Chapter III. Barney Webb was meanwhile organizing all of the other material to make what I consider a superb conference proceeding. It was well worth it; the proceedings have been very heavily utilized, and there has not been another Fermi surface conference.

At the conference Art Kip from Berkeley suggested I visit Berkeley for a stretch. Lucky and I talked about it and decided it would be a good idea. The laboratory agreed to it and we arranged a three-month visit. Joan Kip found us an apartment in El Cerrito, just north of Berkeley and we flew out, in those days by propeller plane with several stops between Albany and California. John, at two, slept on the floor in front of us. Rick was amazed that there was no snow in California, not even piled up at the gas stations. A spur track for the Sana Fe Railroad passed just behind the apartment and John was always the first to hear a train, so he and Rick could run out and wave to the engineer.

I had an office with some theory students and I gave some talks in the department associated with Fermi-surface studies such as with Art Kip's group. My desk was with Charlie Kittel's students, which was fine with me, and Charlie and I became long-term friends. Charlie once brought in a paper he'd been asked to referee and asked me to evaluate it. I suspect he thought I belonged to him. I said no, I wasn't interested in that particular article, and that took care of it. We got on fine.

Lucky and I bought an old Buick at a used car lot that the dealer agreed to buy back at a reduced price when we left. We made a trip to Southern California with it, visiting a couple of physics departments, and returning up the coastal road. The drive along the coast is a little scary, but nice to be on the inside lane, going north.

Lucky and I set aside one evening a week for dinner in San Francisco, with a coffee at Enrico's at the end. One evening, Rusty Hoefle, from a neighboring streetcar to ours at Sages Grove, happened to pass by and joined us. We found some delightful teenage girls to babysit for these excursions. One I remember telling me about fishing for cod with her

Rick and John at the ocean in California.

father and the cod having both eyes on the same side because they swam at the bottom. I didn't believe her at that time. It was a wonderful trip and maybe set the stage for our later return to California. We returned by train, sleeper coach as I remember, and arrived in Albany on John's third birthday.

Ivar finished his course work at RPI and needed to do a thesis. The most sensible thing would have been to write up his tunneling experiments, but RPI didn't want to feel that they were rubberstamping General Electric degrees so he had to do the research for his thesis at RPI. We talked about it and decided that the solution was a theory thesis because you can't tell where it was done. We picked a topic which the theory professor there, Hilbert Huntington, could feel he directed, but I could help with, something about electron scattering in metals. It succeeded, everybody was happy, and Ivar got his degree. I don't think he ever published it.

I thought that the Fermi Surface Conference culminated in the successful understanding of the electronic structure of most metals, and it was time to move on to new problems. However, many people continued in this successful field, measuring the change in Fermi surfaces with applied pressure, and such. I chose not to do that.

The main thing we had learned was that the electrons in metals are almost like free electrons, little affected by the crystal lattice. This was a surprise. Later when I was in Russia, a Professor Kaganov, who was a close associate of the leading Russian theoretical physicist Lev Landau, told me that he, Kaganov, had seen my paper on free-electron Fermi surfaces and thought Landau would be interested. Landau had said: "It's wrong. There is no small parameter. The interaction between electrons and the lattice is strong!" He raised the right question, but had the wrong answer. The strong interaction produced core electron states which could be taken out of the problem and were not important after that. What remained was a weak interaction, which I called a *pseudopotential*. It determined the properties of the metal, including the Fermi surfaces but also including all other properties.

To understand this pseudopotential I studied the theory of electron states in metals by Conyers Herring from 1940. Conyers was now in the theory group at Bell Labs where I had interviewed. His method could give the numbers which characterized the interaction of electrons with the crystal, and played the same role as a weak interaction would. It provided a way to calculate simply the entire range of properties of metals, which I called the pseudopotential method. For example, it predicted the counterpart of the force constants which I had adjusted to fit the measured elastic constants in order to calculate the vibrational frequencies in silicon, in my thesis.

With these calculations of various properties it became appropriate to use a computer. The research lab did not have one at the time (as far as I know the only major industrial laboratory that did not), but one was available at the main plant in downtown Schenectady. It had been seven years or so since I had taken computer programming at Illinois and it was an entirely new world. Programming languages developed by a Dartmouth College professor, rather than machine language, had come into being, notably *Basic*, so his students could learn to compute remotely with no computer on campus. I learned Basic and set out to write a program to construct pseudopotentials, obtain interatomic interactions from them, and vibration spectra in terms of the forces. I confidently took the program downtown to run on the computer. Anyone with the slightest knowledge of computing would have written a small program, and seen if it works, and then moved on to the next step. However, with a few corrections here and there I made it work and published papers on the vibration spectra, etc.

A year or two later a young French physicist, Robert Pick, asked to come and translate my computer program into French Fortran. I agreed and he came, rented a pad somewhere in Schenectady, and sat in the office next to mine, translating. It turned out to be frightening for me. He would continually come in and say: "I don't understand this step." I had never checked anything and was very much afraid each time that he had found a real mistake in the program, but it never happened, and it was very nice having him there. Some years later when we visited him in Paris I met his son, who didn't like me at all, since I had taken his father away for an extended period of time.

This whole project of calculations turned out to be something of a theme for my future work, developing simple ways to understand the electronic structure and properties of solids, and do simple calculations to predict their values. At the time it was simple metals, but I soon went on to the more complex transition metals, and years later back to semiconductors and insulators. At the time I was doing this, other physicists were working out much more complete and accurate calculations of properties with ever larger and more intricate programs, utilizing larger and more powerful computers. That turned out to be the mainstream of activity in the world, but I continued with my simpler approaches.

Bill and Bob arrived in 1962 and we became six. By then we had finished the large upper bedroom and Lucky and I had moved up. We got an extra crib for the third bedroom on the next floor down. At this time Roland Schmitt was living at the end of our street, Cedar Lane. He heard about a large old house on Lake George that was for sale and

In Vermont, 1960. Lucky, her mother Marion, brother Bob, and father Al holding John. Then my mother Gertrude, and Rick down in front.

discussed with me the question of buying it. I strongly encouraged him and he did buy it. When the sale was complete, he had other obligations and couldn't go there, so he suggested that we could go and maybe in exchange clean it up a little. That was a little tricky with our two new babies, but we agreed and arranged for an older baby sitter, Mrs. Walsh, to come with us. We drove to Bolton's Landing on Lake George where we could hire a boat taxi to take us out to this house on the side of Tongue Mountain, the only access to the house. There was a large boathouse on the dock, with an upstairs that even had a player piano. We carefully passed Billy and Bobby from the boat to the dock and ventured up the winding path to the house. The house was huge with very many bedrooms, all equipped with beds. We learned later that for many years it had been serving as an abortion clinic. Abortions were then illegal in the United States and this house was quite remote from the rest of civilization. It hadn't been occupied for some time and was quite in disarray. We did what we could to bring some order and got the kitchen working. It made a nice, if strenuous, vacation, and the water taxi came on schedule to take us back to Bolton Landing.

When Werner Känzig returned to Switzerland, he left us his big old canvas tent and we began camping around the East, initially by Lake Dunnmore in Vermont, not far from Marion and Al. Four of us fit in the tent with sleeping bags and Bill and Bob slept in the car. On one of those occasions the Giaevers came also, I believe camping in the next site. We also did one trip to camp on Cape Cod. I cooked dinners over the camp fireplace, but John seemed to enjoy more the time when we had cooked in the fireplace at home when the power was out for an extended period due to ice storms.

We made use of the sledding hill close behind our house and started Rick and John with skiing lessons. I had done some skiing with the Suits on day trips and Lucky also took some ski lessons. We were often in Vermont staying with the Carleys, and I skied there once or twice. Al was managing the local movie theater. My mother, Gertrude, was a frequent visitor with us and at least once went with us to Vermont. The grandparents were looking forward to coming and helping when the twins came, but in the end they all opted to wait a while before coming.

Our social life in Schenectady was considerable. We had friends from Cornell, as well as those from the lab, and from a bridge group that Lucky had joined. We started having annual New Year's Eve parties. We had a group of four couples with whom we had dinner and bridge once a month or so, Barney and Fran Webb, Jim and Nancy Livingston, and Inger and Ivar Giaever. We initially called it Dinner–Bridge, but later Drink–Dinner–Bridge, and the "Drink" part became more major as time went on, but I think we always managed to play bridge.

My carpool was another social event. We had just one car so it was important to pool, and we had five of us in our housing area, Willowbrook Park, so I needed the car only once a week. One member was actually Bill Johnston with whom I had carpooled from Daisy Lane. Another was Barney Webb, my *Fermi Surface* coeditor. We could discuss world affairs every day and sometimes it was important for work. When I was getting ready to do my pseudopotential calculations, I eventually needed the results of other calculations; calculations of the electronic structure of copper atoms was first on the list. When I went to the lab library to hunt them up I found that they had been carried out by Bill Piper, who was one of our carpool, the only calculation for copper. On another occasion Charlie Brook, another member after Barney left, mentioned that he and his group were just starting on a major program to make Lucalox, a special glass based on aluminum oxide for large flood lights, clear. The current product was blurry but had been getting better. I said: "That's Al_2O_3, isn't it? If so I have an idea that it's birefringent, and then you can't make it clear." (That was a birefringent crystal that I had used for my device at Owens-Illinois Glass Company. It means that in the crystal a light beam divides into two beams of different polarization going in slightly different directions.) Charlie hadn't heard of birefringence but later in the morning he came running to my office with a sample and said it was birefringent. A little inspection indicated that it was about as non-blurry as it could get. They stopped the program and saved the company a very large amount of money. I suspect that it was the largest contribution to General Electric's business that I had made during my stay there. I wrote a short GE internal report on the subject and some years later I got a call when at Stanford from a GE division in Philadelphia that made the domes for the front of missiles. They needed to be transparent to radar. He wondered how my research on blurriness had progressed. It did get me a temporary consulting job and a few trips east from California.

During my last year at GE I wrote my first book, *Pseudopotentials in the Theory of Metals*. I had wanted to do a book earlier on tunneling jointly with Ivar, but he was not ready yet. Writing a book didn't change my life much. We continued to have visitors and seminars and discussions, but most of the time in my office I was typing for the book. I had a plan to keep my interest up: there is an interesting defect in crystal lattices called a stacking fault. For each topic, like electron scattering, I could calculate it for stacking faults. In the final chapter about total energies, I would calculate the energy of the stacking fault and then go look up the experimental value to check. All went on schedule and in spring, I calculated the energy of the stacking fault in magnesium as 50.3 ergs/cm^2. I headed off for the library and found that there had been an estimate for zinc, but not for magnesium. The one for zinc was by Buford Price *who was in our carpool* at that time. Somehow I had never brought up that particular calculation in the carpool. He had speculated 30 ergs/cm^2 for zinc and expected that magnesium would be somewhat higher. From my point of view that was perfect. I didn't want to be tested *too* closely.

I had been wondering a little about my life and General Electric. We were doing basic research there and no one was complaining but I wasn't sure it was a good investment for GE and whether we should be doing something more applied. Others were having similar thoughts and the story about making a "geek" went through our group. Every carnival had a geek that people paid 25 cents to see. He sat in a cage and they'd throw a live chicken in with him, which he would bite into and drink the blood. The carnival was in trouble when the geek died because there were none to hire, so the carnival owner had to *make* a geek. He'd find an old drunk in whatever town they were in and tell him about the geek dying and that they had another ordered from Borneo. In the meantime, they needed someone to sit in the cage. No throwing a chicken in, and he'd have drinks and food to eat, and a place to sleep. The drunk usually agreed, but after a few days the owner would tell him that the crowd was getting restless, and they needed to throw a chicken in the cage, but he should just ignore it. Step-by-step over the next few weeks they had made a geek. And we were already sitting in the cage!

About that time Marvin Chodorow called. He was passing through Schenectady and said he'd like to talk to me. I knew his name from his thesis work before World War II, in which he had calculated something of the band structure of copper. He was the chairman of the newly formed Applied Physics Department at Stanford. They were hoping to hire a solid-state theorist and invited me to visit. I did, and very much liked the people I met there and the general atmosphere at Stanford, which I had never visited before. I received an offer and the time seemed right to me, though Lucky had misgivings about moving our family west and leaving our other family behind. Barney Webb had left GE for Wisconsin and Henry Ehrenreich for Harvard. I accepted the offer.

We took a last trip down to New York City to go to the third wedding of Dave Plant, my Cornell roommate. At the wedding we ran into Julie Maser, whom Dave had dated in my last year at Cornell, and her new husband, Dick Maser. Dick was finishing up his preparation to be a plastic surgeon. After the wedding, we four took the boat tour around Manhattan, and went to Greenwich Village for dinner. We had a delightful time, but when Dick learned we were moving to California he said: "Damn, that's always the way. You meet some new friends you really like, but then you're never going to see them again!" The next year Dick took a position at the Palo Alto Medical Foundation and we remained fast friends as long as Dick and Julie lived.

Later in the spring I had a chance to go to a conference in Arizona, which paid most of my way to California to look for housing. Lucky flew out to join me and we explored what little was available. After a couple of days we gave up and arranged a rental in Los Altos for the summer. We had seen one house on campus which looked good, but it turned out to be unavailable. On the way home, I called the owner of that house from the airport and said we were interested if anything changed. She said actually she *was* willing to sell it, and what was I willing to pay. I said $42,000. She said she was hoping for a little more but she'd take it!

It would be available in September. You can imagine our unbridled joy as we returned to Schenectady.

We had time to say farewell to our many friends in Schenectady and get packed up. They arranged a delightful fairwell party for us. We sold our house, for approximately what we had paid for it, some $20,000, so our California house was a stretch. Stanford was paying for the move, but in the end gave us an adequate lump sum rather than paying the bill. That meant we needed to decide if each item was worth the $1 per pound it would cost to ship it. The board and brick bookcase didn't make it. We were still driving our wedding Chevy, though it wasn't adequate for a family of six. Ivar had just bought a Ford station wagon. I knew he compared all options before choosing and drove a good bargain so I went to the Ford dealer and said I wanted just what Ivar got, in a different color, and at the same price, which we got. I constructed a board barrier between the back seat and the deck to reduce the interactions between kids (two on the seats and two in the back deck) and we set off for the West. We stopped in Ohio on the way through, out Route 66 to LA and Disneyland, and finally up to our new home, in northern California.

The family in 1964.

And what was wrong with the Chinese flatiron perpetual motion machine? The problem was using *point* particles. Real particles of matter must have finite size and with finite size some of the radiation from the left just misses the right particle, more than misses the left particle coming from the right particle, because of the different curvature of the reflector on the two sides. It's amazing that the radiation just balances and remains balanced no matter how small the particles. It's only with mathematical points that it doesn't balance, and point particles don't exist in the real world. The second law of thermodynamics is a law of the physics world. In the world of mathematics, point particles are allowed and *usually* behave like real physical particles.

Chapter V

SETTLING AT STANFORD

We arrived at our temporary quarters at Creston Drive in Los Altos. It was a twenty-minute drive to Stanford, where I had an office in the Microwave Laboratory, headed by Marvin Chodorow. Rick quickly made a friend on the street, from a native California family. When I met the father I took advantage of his California background and asked how you say "Junipero Serra". He replied "Juniper Sierra", but it was nice to know a local. There weren't many natives in the Department at Stanford.

I had been invited to speak at a summer school in Aberdeen, Scotland, and left for three weeks not long after we arrived. Lucky was

John and Walt at Rosatti's in 2018.

busy with the kids and even discovered Rosatti's Road House where they could play at the San Francisquito Creek that ran behind it; Rosatti's is still a haunt of ours.

When I returned we made trips to the coast, which was exciting but generally surprisingly cold. We went to see our new house, which we hardly knew. All surprises seemed to be good, such as sprinkler systems in front and back. We met with the owner, Marjorie Myhill, at a Stanford office to finalize the sale. It was monitored by Ed Scoles and seemed remarkably simple. Nowadays it requires lawyers and inspections and is a major process. Our furniture arrived at the house on September 1 as scheduled, including Mighty Mo, which was a big rock that the boys had treasured at Cedar Lane in Scotia, and we settled in.

There was a paved walk all around the house so the boys could ride their cars around. At that time there were 36 other kids on our *cul de sac* so it wasn't lonely. Lucky took advantage of her teaching experience, and her masters in psychology, and supported artistic and social activities and projects for our boys and the neighborhood. She opened the house to the whole neighborhood of kids. They flew kites, made candles at the beach, colored eggs for Easter, and spent a lot of time writing and drawing. One of the large cabinets in the Play Room was

Walt at the beach, 1968.

known as the "What Shall I Do?" where all the paper, pens, and other craft items were kept. Over these years our sons took art, film, dance, skating, piano, trumpet and trombone lessons. Our family dog, Pokey, a stray who wandered into our backyard one day in 1967 and who we later adopted from the pound, was completely independent and had free reign in the neighborhood. He died in 1984 at the age of 17, having seen all the kids grow up and leave home.

The boys had access to "all the privileges of the Stanford Community" including Frost Amphitheatre, Hoover Tower, Tresidder Union, and later fraternity parties. They would build everything from go-carts to placards for hot dog sales in the garage, which was equipped with a work bench, tools, and plenty of scrap plywood. They held carnivals, built forts, and even started a newsletter for the neighborhood.

I continued with my office in the Microwave Lab. I purchased a bike, the standard three-speed professor's bike of the time, and it was an easy ride to my office. We got by with one car. I continued to bike to the campus and back for fifty years; then it seemed questionable for an 85-year-old man. In a few years, I then bought a three-wheel adult bike and continued. Lucky also had a bike for a while but eventually gave it up. Rick and John were already biking when we moved in, and Bill and Bob soon were. For a while we had a two-person bike, but it was stolen and we bought a second one. It was also eventually stolen. One activity with the tandem bikes, of which I was unaware at the time, was for one rider to close his eyes and the other to ride off to some unfamiliar place and let the first figure out where he was.

The move to California at the beginning must have been a strain because one day I collapsed in my office, alarming others in the building, but I was quickly back to normal. I was given an electronic calculator of my own by the Department, quite a treat for me. It was bigger than a typewriter and could silently add, subtract, multiply, and divide, and had a brand new feature that it could perform square roots! I was used to electric ones that were mechanical, spinning gears, and square roots were a complicated procedure. Before many years, the Department gave me an Apple Macintosh lunch-box computer. Things were changing rapidly.

It was arranged for me to teach the Solid State Theory course in the fall. I had never taught a course in my life, partly because of my fellowship in graduate school, but I had given many physics talks and I proceeded to organize a year-long course. I began with group theory, the useful part that I had taught to our office group at Illinois. I was pleased on the first day of class that there were too many students to fit in the room and we needed to move to a bigger one.

The Department provided me with a grader for the course, an English student named Chris Wilkenson. That was very fortunate because he knew a lot more about teaching than I did. Instead of writing on a black board as when I was a student, it was usual to make transparencies, eight-by-eleven inch transparent sheets that you then laid on a projector. That made it possible to organize each lecture ahead of time and I could watch the class as I talked to see whether they were understanding. Chris also sat in on the classes and made very helpful suggestions about the lectures. I gave a test in the middle of each quarter, and a take-home final at the end of each and consulted with Chris about those. Also when a trip came up and I wanted to miss a class, I could leave it to Chris with confidence. Chris turned out to be a very good professor teacher!

For the second year I had the transparencies, and therefore the organization of the course to start with. It was the main graduate course in solid-state physics at Stanford, and I taught it for perhaps fifteen years. It was natural to write it up as I progressed and it became my book, *Solid State Physics*, published in 1971, with problems at the end of chapters and all the trimmings. I wanted the text to be in the McGraw-Hill International Series, partly because these were the off-green-colored series of many of the texts I had learned from in college and graduate school. That was easy because Leonard Schiff who was the editor of the series was in the Physics Department at Stanford. Some time along I sent a note to Bill Benjamin, the publisher of *Pseudopotentials*. I don't remember what it was about, but I remember his one-sentence letter back: "I understand you now publish with McGraw Hill." I hadn't realized he considered me *his* author. My professor, Fred Seitz, had written a text *Modern Theory of Solids,* which was the standard text, and I think mine replaced it for a period. Charles Kittel at Berkeley had written an undergraduate text which was quite standard, but it couldn't use quantum mechanics for an undergraduate course so it wasn't suitable for a graduate text. He tried a graduate text but it wasn't any competition. At a later date, Neil Ashcroft and David Merman wrote a solid-state text. The publisher sent me a copy of a draft of the first part for comments. It was interesting to look at. The style was very leisurely and in the first two hundred pages they covered what I would in fifty. I said that to the publisher, but noted that maybe it could be better for some. Readers may like to skim along and pick up a little on the way, which they could with Ashcroft and Merman, but with mine it wouldn't work. Mine was concentrated enough that a reader had to stop and absorb the idea presented before proceeding. My book worked carefully through all the parts of solid-state physics and culminated in the Bardeen–Cooper–Schrieffer (BCS) theory of superconductivity. Ashcroft and Merman never got to that crown jewel of solid-state physics and their text is considerably longer than mine. I think in the end theirs was more popular.

There was an Army Research Program (ARPA) at Stanford, administered by Bob Huggins, which provided me money for summer salary and for support of a post-doc and a couple of students. It also actually paid half of my academic salary. I hadn't heard about that when I accepted the offer and years later when ARPA started to back off I didn't worry about it. When I heard it would no longer pay half my salary I went to the chairman, then Hu Heffner (not to be confused with Hugh Hefner of Playboy fame) and said "I think *you* have a problem." He smiled and said *he* did have the problem. The University made up the difference.

My friend from RCA, Frank Herman, had since joined the Lockheed laboratory in Palo Alto and hired me as a consultant to Lockheed. It was considered acceptable at the university to spend a day or so a week in consulting and I did something like that. Much of it was working out problems related to Frank's interests. I enjoyed it and it made a substantial contribution to our income. Consulting was to be an enjoyable and broadening part of my life throughout my career at Stanford.

Soon after coming to Stanford we met Joel and Fran Smith at a party given by Bob Beyers, a Cornell friend, and actually one of the group who took over the radio station in my senior year at Cornell. Joel was then Dean of Students at Stanford. The Smiths asked us about sharing a house for a week in the summer at Fallen Leaf Lake. We hadn't done that sort of sharing before, but we agreed and it was the beginning of our very long relationship with Fallen Leaf Lake and with the Smiths. The house had been built by the Hinds and Botsford a few years earlier. The arrangement of the house, the boathouse, and the path down, were remarkably like the house we had stayed in at Lake George on Tongue Mountain, but without the extra beds, and this one had a modern A-frame style.

The Smiths had two girls, Becky and Jennifer, similar ages to Rick and John. We all got on well and even had access to the waterski boat that belonged to the house. We all had a chance to learn waterskiing. We repeated the stay with the Smiths a number of times, until their lives got complicated.

After that the boys invited friends to stay with us and we usually had a pretty full house, often with many of them sleeping on the deck.

The house at Fallen Leaf Lake, seen from the lake.

Bob, Bill, John, and Rick at Fallen Leaf Lake.

Sometimes friends would take the plane up to Tahoe airport and we could pick them up, sometimes they came by bus or driving. The main activity was sitting on the dock and swimming, but we took hikes, sometimes to Floating Island Lake, which *did* have an island which drifted from one end to the other depending upon the wind, or to the top of Mount Tallac, behind the house with the white cross of snow. We had the use of a Sailfish, a small sailboat, a canoe and a boat with an outboard. We would sometimes all go to the Diving Rocks, down the lake, for some higher diving. We also played cards, most often *Oh Hell*, as I remember. Occasionally there was a trip to State Line and the casinos, but that was not the norm. There was always beer in the refrigerator, but drinking was not a major part of the activity.

There *was* a period when the house was not available, I forget why not, and we rented the house of John Kraft, down the lake from us. John Kraft is actually the one I remember particularly for teaching the boys to waterski, using his own boat. We also became quite close to Bill and Sally Marriner who had the house next to John Kraft. Eventually our regular house became available again and we returned every summer, until finally a few years ago the wife of the present owner, Chris Botsford, retired and they spend all their summers there.

Once we had moved into our San Francisco Court house, Lucky organized the annual 4th of July parade. There had been one on Cedar

Lane in Schenectady and we reproduced it at Stanford. The kids decorated their bikes with colorful crate paper and rode down San Francisco Court, behind a flag and crepe paper ribbon, into San Francisco Terrace and out, down to the end of the street and back to our driveway for refreshments. This activity has continued for over more than 50 years, drawing entrants from around the campus. For a short time, when our sons were too old to march, the crowd thinned out and one year we adults marched, while some teenagers watched, but after that it increased again and stabilized. There was one year when Channel 4 sent a camera crew and we had a three-minute piece on the evening news. Another year, we were away and Marlene Handy kept it going.

Lucky also organized an annual Easter-egg contest on the street, with big egg-shaped papers that people decorated. We pinned them on our garage door to be judged by a suitable figure from the neighborhood. She arranged ice skating parties at the Winter Club in Palo Alto. Each year we'd reserve the Club, our boys, and we, would invite friends who would pay for their skate rental, and we provided hot dogs and refreshments. At some point it got large enough that we rented a bus to travel to and from our neighborhood. Later, as the crowd dwindled, we continued, but joined a regular open session. We still provided the hot dogs and refreshments.

Early on there were visitors, Leigh and Marilyn Browder, on sabbatical and living in the Huggins' house across the street. Marilyn suggested to Lucky to form a group of four couples to trade gourmet dinners. We added the Masers and Bill and Barbara Ingram and started going through a Gourmet Cookbook, each family taking a turn hosting and preparing the main course and others bringing the extras. When the Browders left, the remaining three couples continued, with the host inviting a guest couple. Eventually one of the guest couples, Bob and Betty Hilmer, became permanent and we continued, meeting every three months or so. One time, as a group, we hired a lighthouse on East Brothers Island in the San Francisco Bay for a night, a possibility that the Hilmers knew about. A boat carried us out and back from Richmond and the couple living in the lighthouse provided dinner. It was enchanting and Lucky got the idea of giving her Uncle Willis's silverware

The lighthouse on Little Brother Island.

set to the lighthouse to dress up the dinners, which we did. It was even deductible! I see on the web that it is still available for rental and I expect they're still using Willis's silverware. The group continued for many years, until Bill Ingram died, and Barbara then invited a guest. Next Julie Maser died, then Bob Hilmer, and then Dick Maser. We added Ann Clark who had been a guest and it still continues in a less regular way. It has been a mainstay of our social life.

It is customary to apply for grants to support students and post docs so when we came to Stanford, I applied to NSF, the National Science Foundation, for a such a grant and received one the next year, making a smooth transition out of the ARPA program. My first post doc was a Nigerian PhD student who was with Volker Heine at Cambridge. I knew and liked his work on pseudopotentials, but Volker was a little surprised that I didn't discuss it with him. I think he would have cautioned me that this student might not be easy to deal with. He arrived with his wife and son and was given an apartment in Escondido Village. We started a project together and discussed it as we advanced, and he seemed to be making good progress. We were writing a paper on it together when in discussion we found a mistake he had made. It was not serious but required redoing the calculations and he agreed to do that. However, when he finished redoing all the final numbers, he calculated them to be exactly the same. When I pointed it out he said the correction didn't change the answers. I said, "Now really, that can't be." He said if I didn't believe him I could do the calculations myself. So, I took my name off the paper and he proceeded as he wished. His life continued to be tumultuous, with some ruckus concerning parking in Escondido Village, and then concerning his move to North Carolina for his next post doc position. He ended up in a big academic position in Nigeria and I feel for those students who have had to deal with him. Years later I heard him give a talk on nuclear physics at an APS meeting, which struck me as utter nonsense, but we had a nice friendly talk after.

My second post doc was from Dartmouth and a very nice and easy-going guy, but he was not much interested in the project we started together. He ended up publishing something with another professor here before he left. The third post doc was much more successful, Sokrates Pantelides, a Greek Cypriot from the University of Illinois. He was enthusiastic, hardworking, and we worked and published together. One of my graduate students at the time was Salim Ciraçi, from Turkey. Sok worked with him for sometime, but when Turkey invaded Cyprus it became no longer possible. Sok and I did a number of studies together on compound semiconductors like zinc selenide and on silicon dioxide, the major component of glass. Sok went on to IBM and I spent time with him there. He ended up as a professor at Vanderbilt University in Nashville. My grants kept being renewed but they didn't keep up with inflation so after Sok, I didn't have post docs, but visitors who provided their own support.

I generally had one or two graduate students working on problems with me. When another inquired, I asked the business manager at Ginzton if I had enough funds in the account and he always said yes. Then one day he came to me and said I needed to raise more money, so I backed off on the number. I guess he thought saying yes was being nice. I've lost touch with many of these students, but in looking back I see that their thesis topics mark the evolution of my research over the years.

Bob Shaw was one of my earliest ones. He was the son of the astronomy professor at Cornell whose course I had taken. The father was a real problem and I thought his course was bad. At that time, he was lecturing about the difference in the Dopler shift depending on whether the observer was moving or the source was moving. There *is* a difference for sound waves, but the main message of special relativity is that there is no difference for light waves. In other courses, I would question it in class, but I knew he was paranoid, so I waited after class until everyone else was gone and said I wanted to raise a question. As soon as he detected that I was going to question something he'd said, he exploded. I apologized and backed off. So Bob and I didn't discuss his father. Bob did his thesis on an alternative to pseudopotentials, called model potentials, introduced by Volker Heine. Bob went on to Bell Labs, the Cavendish Laboratory in Cambridge with Volker and then became a successful venture capitalist. I see him when business takes him to this area.

John Wills came a little later and we extended pseudopotential theory to transition metals, which was a major extension. He went on to the University of West Virginia as a post doc, when I visited him for a few days. One evening I was free and went to a local pub. It was crowded but no one seemed at all friendly. I decided to finish my beer and go back, when I struck up a conversation with a couple of local young people and got a feeling for the local culture. They talked about how some locals go away for school and you think they'll escape Morgantown, but then they get their girl pregnant and return, get married and never leave. Mining was the main business in town and it wasn't doing well.

Hu Heffner was part of the founding of our department, but left to be among the White House staff in Washington before I came. When he returned we became good friends. The new Faculty Club didn't have a liquor license and we shared lockers in the bar area to keep a bottle. I shared a locker with Hu. Then the Club applied for a liquor license and closed the lockers. While waiting for the license, a group of faculty started meeting for lunch in the area that had been the bar, behind closed doors, and brought in jugs of wine from outside. Hu was a part of this group and invited me to join in. It evolved into the lunch group that I joined, with varying frequency, ever since. By now it is more often referred to as the Old Farts. It seemed to me that Hu had been away from Stanford too long to get back into the swing of research and he took various administrative positions in the university, including chairman of our department. Suddenly Hu became terminally ill and died, I would say in his forties.

This lunch group at the faculty club was a window to a new world for me. I didn't like having wine with lunch so initially I only went once a month or so, but I then realized I could go and *not* have wine and eventually I pretty much went whenever I didn't have anything else to do. It remained in the same room in the back of the Faculty Club by the bar, with a beautiful quote carved from wood high on one wall. The quote came from the old Faculty Club about the time I arrived at Stanford and read "Knaster den Gelben hat uns Apollo präpariert." I think it's from an old college drinking song and translates roughly as "Apollo prepared the golden tobacco for us." Actually I understand that Apollo was a make of pipe tobacco and the reference was to it.

The group that met was varied and eventually grew to fifteen or twenty regularly and forty or fifty on Fridays. By now it had dwindled down to four to six on Fridays and almost no one during the week. The remaining group included Jim Collman, a retired chemistry professor and the only political conservative in the group. It's probably good for us to hear the other side sometimes. Another is Clay Bates, retired from Materials Science, and our neighbor Bob Huggins, also from Materials Science. Then there is Art Barnes, well known as director of the Stanford Marching Band because at each football game they would announce the band playing the national anthem, "arranged and directed by Arthur P. Barnes". There is also the libertarian Tom Moore from the Hoover Institute who would bring his wife Casandra, and sometimes Alan Grundmann whom I knew as administrator in the Physics Department but who became the manager of Stanford's nature preserve. Also Jim Harris from Electrical Engineering and who was a student in the first class I had taught at Stanford. I used to be about the youngest member of the group. As time went on the older ones died but almost no new younger people joined so I remained at the young end.

The earlier group made an interesting collection of friends I might not have known otherwise. A central figure was Ron Bracewell, from Electrical Engineering but who considered himself a physicist. He had extraordinarily broad knowledge so whenever something came up and we didn't know the facts we would call on Ron, just as one Googles now. I noticed that on occasion when he didn't know the facts he switched to the Socratic method and started asking questions. There was Harry Mosher, from Chemistry, who later carved a new slogan, *Edite, Bibite, Collegiales* in Latin rather than German but from the same drinking song, and mounted it on the opposite wall. Another member seemed to me to have the worst judgment of anyone I had ever met. He regularly bet in a football pool that I thought was a guaranteed loser. He thought that stock prices were going to plummet so it was time to sell. Only one person in our group took his advice and shifted his retirement funds into bonds, so he retired with very much less than the rest of us. The man whose judgment I questioned discovered that he could buy a huge house in Oregon at the price his modest home here was valued, so he sold his house and moved up. After a year he decided he didn't like living in Oregon, sold his house and came back when prices had doubled in the year. He bought a house high on a hill in Santa Cruz with the only access by a long staircase from the road for himself and his aging wife. There was also Pierre Noyes, a physics professor from SLAC who seemed to me to have similar judgment. He came to lunch one day all aggrieved about the Shah of Iran. I didn't know anything about the Shah but I thought if Pierre was so against him the Shah must have some virtues. Within days Ayatollah Khomeini led the Islamic revolution which overthrew the Shah, putting Iran further back in the dark ages. The Shah at least had the virtue of being better than the alternative.

David Redfield was a friend in Materials Science who asked me sometime about this group he had seen enjoying themselves at lunch. I didn't really think he would fit but I told him about it and invited him to try. He asked if anyone smoked there and I said only Joe Judd who occasionally comes and smokes his pipe. That ruled it out for David but after Joe died David came and became an integral part of the group. I met Jim Stockdale at dinner with neighbors and invited him to lunch. He had been shot down in his plane over Vietnam and spent seven years as a captive in North Korea. He came back to the lunch off and on and was a lively addition. He eventually became a vice presidential candidate when Ross Perrot ran for president of the country. There was Joe Ruetz who was head of the Stanford athletics program and Sandy Dornbusch from Political Science. Sandy was among those who sometimes held forth at some length at lunch. One day when Sandy said: "I have to leave in a minute, but I have a story I want to tell first." I was pleased that Alan Grundmann said: "No, Sandy, if you tell a story, you have to wait till someone else tells a story before you leave." Doug Skoog had become very prosperous by publishing a popular undergraduate chemistry text. Chet Berry was in charge of both the Tresidder Union and the Faculty Club. Dick Blois from the legal office became our family lawyer and Kirk Roberts from Chemistry had lived down the road from Lucky's parents on Cider Mill Road in Vermont. It was a group with diverse connections.

When Hu Heffner died, his student, Dick Meserve, came to me. Dick was more like a post doc than a student and the only student with whom I ate lunch regularly, at Tresidder Union. At the time he worked with me, I was thinking beyond pseudopotentials, toward semiconductors, but with the aim of providing a basic theory of the properties such as pseudopotentials had provided for metals. The clue for metals was that the effect of the potentials was small compared to the energy of motion of the electrons, and could be treated as a correction. In semiconductors the opposite seemed to be true. Maybe we should start with electrons bound to their individual atoms and add their motion from atom to atom as a correction. This description was not new, but was an old view called "tight-binding" theory, considered not accurate enough to be useful.

We would try to improve upon it and Dick Meserve was part of that effort. His thesis developed corrections to the original tight-binding theory. When he was finishing his degree I urged John Bardeen to take him on as a post doc, but John ended up picking someone else; John later agreed he'd made a mistake. So Dick stayed on with me as post doc till he could find a suitable job. Times were not good for jobs and he had applied to Harvard law school as a backup. Then the acceptance came through, and he figured why not. He finished his law degree, was a legal trainee in a prominent Boston law firm, and then joined the White House scientific advisory committee to the president, as I remember the only member to be both a PhD and a J.D. When that job ended he joined a law firm in Washington. Again he was the only one with a PhD and a J.D. Before long he was appointed head of the United States Nuclear Regulatory Commission. After that he became president of the Carnegie Foundation, which happened to have its annual meeting of the board at Stanford. He saw to it that Lucky and I were invited to the banquet for those meetings. It was a remarkable group to have dinner with. I remember at one dinner meeting a Will Hearst at our table, who we soon realized was the grandson (William the third) of William Randolph Hearst, the subject of the movie, Citizen Kane. Sitting next to me was Charles Townes from Berkeley, co-inventor of the laser. It was a little different crowd than we were accustomed to. Dick is now retired from that position and is consultant to a company trying to make fusion power, similar to my current position but it sounds like a longer shot.

Another graduate student around the same time was Maarten Heyn, from Holland and studying to be a biophysicist. He was homework grader for my solid-state course and got interested in our problem. He was remarkably able and I kept encouraging him to go back to biophysics but, as I have learned, Dutch students sometimes have very strong convictions about what they should do. For his thesis, he proved something that interested me, but I'm not sure it interested anyone else. He got his degree, did a post doc in biophysics in Switzerland and ended up as a Professor of Biophysics in the Freie Universität in Berlin. There are occasions when he comes to Stanford and I see him then. He even came with his wife once to visit us at the beach. Another Dutch student I dealt with was interested in astrophysics. He told me about his theory of sun spots which sounded interesting. We talked about it and thought of a way to test it. He found the necessary experimental information and unfortunately it indicated that his theory was not correct. I said that was too bad, but it happens and he could go on to something else. He said absolutely not, this was his theory and he would stick to it, which he did. He even insisted that I serve on his examining committee. After his talk I had to say that I didn't find it convincing, but the others were astrophysicists and I was not. He passed the exam and I have not heard of him since.

Howard Mises was a student I knew who had taken my course. At some time, he came to my office, devastated because he had been fired by his professor, Cal Quate. I never did hear why, but Cal was hard to figure. I said Howard could continue with me on his problem until he finished, which he did in a year. He was doing experiments which provided pictures of semiconductor surfaces, some of which were quite unexpected. He found a very clever way to demonstrate the origin of these unexpected results using the transparency projectors one used for talks. Doing this at a meeting on the subject brought him a standing ovation. Another, Clinton Cooney, was less successful. He had been a chemistry graduate student for a long time but with no success in research. I gave him a problem which seemed adequate to be completed in a year. He struggled through it and got his degree, but he was equally unsuccessful holding down a job. He was a burden for several years.

Salim Ciraçi, the Turkish student I mentioned in connection with Sok Pantelides, came a little later. He was a student in the Materials Science Department, and had taken my course. By this time I had stopped making corrections to the tight-binding theory and was simplifying it to calculate properties, what I called a *bond-orbital model*. Salim's thesis continued in that vein. He ultimately returned to Turkey, to Bilkent University in Ankara. I was invited there when he was retiring as Dean of Engineering, so Lucky and I visited Salim and his wife Ülkü. They took us to a Turkish restaurant that he thought I would like, and it was really the best döner kebob I ever tasted.

At the entrance to the university there was a large statue of the founder, and when they were honoring Salim for his work as dean, this founder was also there to publicly thank him, holding Salim's hand the entire time. Salim looked a little embarrassed. We took the chance to visit Istanbul for a second time. The first time we had been there we ate at a marvelous restaurant in an underground water cistern. We didn't know where it was so we asked the desk clerk when we checked into the hotel. It was a part of this same hotel! We definitely ate there again. We also went to the famous Istanbul bazaar, like the casbah of Algeria. On the way in, we were addressed by a Turkish man who insisted on showing us around. It turned out to be a very nice way to do it. We bought him lunch and eventually he took us to his small room in the bazaar where he sold carpets. We weren't in the market for carpets, but I think we bought some pillow covers.

Soon after Salim left Stanford, a graduate student from the Physics Department joined me for a quarter. I gave her a small problem to get started. It turned out that she didn't get very far and she wanted to refund her pay. Of course not, but she moved on. Next came a Norwegian student, Sverre Froyen. When we first met I gave him the same problem. He was back in a week or so wanting to talk about some other problem. I asked him about the one I'd given him and he said: "Oh, yes, here is the solution." He was quite remarkable. Together we greatly simplified and improved the theory of transition metals. Even the values of the important parameters came out of the theory. One day when Sverre came to my office, I asked him: "Didn't you used to have a beard?" He replied: "This is the first time you've seen me without one." I guess I don't get an "A" for observation, but maybe a "C".

For all of these tight-binding calculations such parameters were needed to describe the coupling between orbitals on neighboring atoms. For semiconductors I had gotten them by studying the results of full band calculations for individual materials, looking for the systematics, and fitting the parameters for these. There were four numbers and I had them posted on my office wall. Then when the corresponding numbers for the much more complicated transition metals fell out of the calculations with Sverre, I thought there must also be a simple way to get them for the semiconductors. Almost immediately, I realized that there was: we had to recognize that not only was this tight-binding view correct for semiconductors, but the free-electron view we had used for metals also applied to semiconductors. I told Sverre and went off to listen to a seminar, but I couldn't stand to wait. During the seminar I worked out the first of the four numbers on my wall, -1.42, and got $-9\pi^2/64$, which equals -1.39, almost identical! The other three were similar. It may be hard to imagine how exciting that was to have something so vague and approximate become so clear and exact. We published the advances together and it essentially completed the effort to have simple representations of all of the usual crystalline solids.

Sverre went on to do a post doc with Marvin Cohen, a computational physicist at Berkeley. After a few months, Marvin told me that Sverre did just what you hope a post doc *won't* do, spend his whole post doc writing a new program rather than producing results. Marvin wanted Sverre to take one of their programs and run it for other materials, but Sverre said he couldn't understand the programs and needed to start over. I was delighted because I wasn't at all sure Marvin's group's programs were good. However, Sverre finished in six months, and reran the materials the group had done before, silicon and germanium, finding the same results, the first time that I really believed them. Then he went on to the new materials such as gallium arsenide, which I could also believe. Sverre went on to the Solar Energy Research Institute, SERI, in Golden, Colorado, under a somewhat difficult but very able Israeli physicist named Alex Zunger. I was relieved when Alex stepped aside and was no longer his boss. The last time I saw Sverre was when I gave a talk at the Colorado School of Mines in Golden and he was in the audience. He had left SERI to become a business programmer. I expect he was very good at it.

The question came up in the Applied Physics Department about my becoming chairman. The first time I said no, I was too busy as scoutmaster, as well as teaching, and I couldn't take on another responsibility. Marvin Chodorow couldn't understand how such side activities could take precedence, but that is the way it was. A few years later when I was not scoutmaster I did agree. I had watched previous chairmen and noticed how frazzled they were, putting off decisions as long as possible. I figured that was the problem, and the best thing to do was to always make a decision promptly. I have to admit that initially it was very trying, but I think I was right. After that things got easier and just moved along smoothly. Our department may have been much easier than most. Our faculty were busy carrying out their own research and lives and asked for little but to be left to just do that. There *were* one or two who were anxious to expand the department in their specialty, and in the end they moved to the Physics Department. I think it was better for all, but it is not usual for a chairman to allow a reduction in the size of his department. I didn't have an agenda nor a desire to set new goals, but we did hire some new faculty. One thing I *did* learn when we had problems to fit everyone into the offices available: it is possible to move two retired professors with their own offices into a third shared office. It is not possible to move one retired professor into an office already occupied by another retired professor. In hindsight that is very logical.

There came a time when it was appropriate for some of the younger faculty to teach solid-state physics, Z. X. Shen, in particular. Z. X. had come from China as a student. He was not admitted to Stanford, but went to Rutgers. After a year he wanted to transfer to Stanford. I was chairman for admissions for our department and was reluctant but he argued that he wanted to work with Bill Spicer and SLAC and the facilities were not available at Rutgers and I gave in. It was clear from the start that he was very able and he stayed on as a post doc after he got his PhD. He then was appointed as Assistant Professor and is now part of the regular faculty in Applied Physics. The first year teaching solid state, he taught one quarter and I taught another. It was interesting that his approach was very different from mine. On any subject, say conductivity and transport properties, his first question would be how do you measure it, and his teaching would follow that line. My first question would be how do you calculate it, and my teaching would follow that line. He used a text by Ibach and Lüth, German friends of mine, and I of course used my book. I thought it made an excellent combination, and we proceeded that way for a while.

During quarters in which I was not teaching solid state, I started on a new course on the electronic structure of solids. As always, I wrote up my notes as a possible manuscript for a book. In contrast to the general solid-state-physics course, this was based primarily on my own research, and the picture was continually changing. Now with each development in the physics I needed to scramble to clean up the manuscript to take the newer view. When I finally decided to publish it as *Electronic Structure and the Properties of Solids*, I chose W. H. Freeman, who had published Linus Pauling's *The Nature of the Chemical Bond* and I had a subtitle, *The Physics of the Chemical Bond*. When I had a completed manuscript, Freeman sent a copy to Pauling and he didn't much like it. I think he went in with a positive outlook thinking he might find something useful to add to his extensive knowledge. The message of the book was more: It's time to start all over at the beginning and do it right. That didn't help Pauling a bit.

Freeman had an excellent copy editor who did lots of rewriting and I struggled to see that it was still all correct. I wasn't 100% successful. The copy editor told me it was his last job; the company was not going to do that kind of editing in the future. It made a fine-looking book in the end, but developments in the understanding continued to happen and I was very happy that I had a chance to do it all over again eighteen years later when I published *Elementary Electronic Structure*. Now I am a little puzzled by the result. There is an organization which seems to tabulate references to scientific publications and send the authors accounts of references to their work. I receive these and don't follow them closely but have the impression that currently there are many more references to the old *Electronic Structure and the Properties of Solids* than to *Elementary Electronic Structure*, fifteen years newer and I think far superior.

I designed a cover for *Elementary Electronic Structure*, which was an array of diagrams representing silicon two-electron bonds, three hundred of them. Each was an oval with a small circle at each end. The one at the center was red, the four neighboring bonds were blue, the twelve next-nearest neighbors were black, and the rest green. After it was published, our son John noticed that one of the little circles that was supposed to be blue was actually green, but nobody else had noticed it. When I published the second edition, I tried with the publishers to have it corrected. They ended up omitting all of the outside bonds which had been green, including only the red, blue and black bonds, and the little circle which had been green by mistake was now blue, but lighter than the rest. I still like to think of the flaw as the false stitch in the Navajo woman's rug, as was the B in high-school English I mentioned in Chapter I.

About the time I published the first book, President Nixon normalized relations with China and it was arranged to have some forty Chinese scientists come for a stay in the United States. One of them, Ren Shang-Yuan (Jen was his family name, but it sounded like Ren so I suggested he change it and it stuck), asked to come with me and I was delighted. He had gotten a PhD just before the cultural revolution and had spent years as a farm hand after. Intellectuals were not in good repute. He became part of our program and contributed, including a paper which we published together. However, he was looking to his future in America and a longer-term situation than I was offering. He brought his wife, Wei Min, who had scarcely left her village before that. I drove them down to Point Lobos to see the ocean and don't remember ever seeing such a spectacular set of waves crashing on the rocks and sending up spray. She didn't know at all what to expect so she wasn't so impressed. After a year or so Shang Yuan moved on to Kansas City, where he could pick up financial support and work his way into the system. Later he brought his sister, Shang Fen, who went to school in the United States, finally with a PhD in physics and ended up teaching in a college in Ohio. She later spent part of a sabbatical with me. I think Shang Yuan returned for some time to China and his home university, but retired to California near a son living here.

We had a visit from Al Joseph of Rockwell International Corporation, whom I had known in connection with Fermi-surface studies. He was looking for joint government contracts between Stanford and Rockwell. I indicated that that wasn't interesting to me, but if he had consulting possibilities I would be interested. The consulting at Lockheed was diminishing at the time. He thought a moment and said: "I'll tell you what I'll do. I'll give you five thousand dollars for a year. Don't be sending time sheets and telling how much time you spent. At the end of the year if we got our money's worth, you'll be on for another year. If not, you're fired and I'll tell all your friends and mine that you cheated me out of five thousand dollars." I didn't hesitate to say "You're on!" and I was soon visiting at the Rockwell International Science Center in Thousand Oaks.

Al was a somewhat explosive character but by the time I arrived at Rockwell he had moved on to other things and I was dealing with Ed Kraut and Ron Grant at the lab and we got on beautifully. Ron was directing a series of interesting experiments on semiconductor interfaces and I did the theory along with Ed, which led to many significant joint papers. Even now with my current consulting, these papers keep coming up. I would visit Rockwell several times a year, leaving after class to fly down and spend part of the afternoon with them at the lab. Then we'd all go out to dinner and I'd come in and spend the next day before flying back that night. We all enjoyed the experience and it went on for many years.

Volker Heine wrote to me that he would be away from Cambridge for a period and wondered if I could come and look after his extensive group in the Cavendish Laboratory. Lucky and I discussed it and thought maybe an early sabbatical would be a good idea, fifth year, 1969, rather than seventh. Pretty soon our boys would be in high school and they wouldn't want to be away. I applied for, and received, a Guggenheim Fellowship which covered the expenses and Volker arranged an apartment for us in the new Clare Hall. It would be Clare Hall's third year and our apartment had been occupied the year before by Ivar Giaever and his family, the year before that by a friend of ours from Berkeley. The head of Clare Hall was my old friend, Brian Pippard, so it was familiar territory. We rented our Stanford house out for the first time and flew to England at the beginning of September. Rick was thirteen, John was eleven, and Bill and Bob were eight.

Ivar had hoped to sell us the American car he had been driving in England and some bikes. I definitely wanted a car with the wheel on the British side and didn't take the car, but we did take the bikes. When we got there I bought a used car, small but big enough to fit the family, an English car which we sold when we left. We were a little surprised when we moved into the apartment that it wasn't as clean as we would expect Inger to leave it, but soon found out that another family had occupied it for the summer after the Giaevers left. We were unable to get the boys into the Pearse School that the Giaevers had used but found Shrubbery School, a reasonable walk down Grange Road from Clare Hall. It was a small private school and turned out to be perfect. Our boys rebelled at the idea of wearing the customary short-pants uniforms, but Shrubbery granted us an exception and they acquired long gray pants and shirts and ties of the required color. Later at a convocation at the school the headmaster said they were proud to be an international school, 6% of the students were from other countries. I quickly calculated that our boys were the 6%.

For me it was a nice bike ride to the Cavendish Laboratory which was then on Free School Lane in the heart of Cambridge. (I had always pictured Free School Lane as out in the rural countryside.) I sat in Volker's office and very much liked his students. I wasn't much impressed with the problems they were working on but I may have been able to help them out with some. I had many friends in the Cavendish Laboratory. There was Nevill Mott, the Cavendish Professor and head of the laboratory, David Shoenberg, and Brian Pippard, all of whom I knew from Fermi-surface times. There was Sandy Fetter and young Bob White from Stanford, and Phil Anderson, half-time at Bell and half-time in Cambridge, and a collection of younger British workers, including John Ingelsfield and Brian Josephson. There was also a Hungarian visitor from Budapest, Belos Vasvari, and a young physicist from Sweden. There was coffee in the morning and tea in the afternoon, many seminars, and lively interactions. During our stay Brian Josephson invited me to a feast at his college. These are quite a big deal at Cambridge so it was an honor to be invited. I needed to rent a tuxedo and it *was* an impressive candle-lit event, including a visit to his quarters in the college.

Volker had been invited to a conference in East Germany, but declined to go and sent John Ingelsfield in his place. When John returned we heard all about it. He was impressed with how bad everything was there, little food available in the markets, and a depressing atmosphere. It sounded exciting to me so I found out about Paul Ziesche, who had organized the school, and began inquiries about the possibility of going. In the end the inquiries bore fruit.

It was an idyllic time for our family in Cambridge. Each weekend we went punting on the Cam River, afraid that it would be the last time for the fall, but it was kept open. Our usual punt trip was a mile or so to Granchester, where we pulled up to the lawn of a pub where we could have lunch. It's common for the punt pole to stick in the mud and pull the punter over, but I never fell. Rick and John occasionally punted, and they never fell either. Lucky and I took the train every week or two to London for dinner with Jack and Annette Blakely from Cornell, who

Walt, punting on the Cam, with Lucky in the foreground.

lived in Clare Hall up the path from our apartment. Occasionally we took the boys and sometimes experienced London theater. Lucky and I had a "local", a pub near Shrubbery School, to which we could walk in the evening.

Directly across the walk from us lived Rudi Dutschke and his wife Gretchen and son, Che Gueverra Dutschke. The boys were fond of Rudi. We learned that Rudi had been a well-known revolutionary student in Germany and was shot by an attacker who hunted him down. He survived, England granted him asylum, and he came to Cambridge. He was well known to revolutionary students and many came to Cambridge to meet him. We would see them in the courtyard, with big hugs, and they would retire to the dining table at the window across from us. I suppose that the visitors expected a place to stay, or a meal, but Rudi was not able to provide either. Eventually one of the visitors would go off to the market and come back with something to eat, which they would all share. Gretchen wrote a book, *The Flying Bus*, I think, and she had all the neighborhood kids over for a reading.

One day near Christmas it snowed and our family made a snowman in the courtyard. A little later Rudi came out with Che; Gretchen took a picture of us all. Then Rudi decided to build an igloo with Che. He knocked down the snowman and used the pieces as a start. He rolled another ball but then got tired and went back to his apartment. I took it as an allegory of revolutionary movements, but some of my sons didn't think that was fair. At some point there was a political ruckus in England and some people wanted to get rid of the Home Secretary. They picked Rudi, who had been admitted to Britain by the Home Secretary, as a reason to call for a vote of confidence in the parliament. The Home Secretary lost and Rudi had to leave. It had television coverage and they even interviewed our John on TV. John said he thought Rudi was a neat guy.

Our snowman at Clare hall. Rudi Dutschke is kneeling in the front. In the back row, John, Corinna Pippard and Rick on the back right. I thought two of the others were Bill and Bob, but they say no.

Our family took many trips in our car around England. One included Stonehenge and in those days you could go right up to it. Rick was checking between the rocks to find the secret to their construction. Our hotel in neighboring Salisbury was very old, with conspicuous fluctuations in the floor levels, but they served a good dinner and we slept well. We also did many rubbings of bronzes in different cities. We made one trip to Europe, by train to the dock and then by small ship to France. The waves were *not* small and I was seasick, but the rest of the family did better. I remember going to a small restaurant in Paris and marching in with the four boys and the proprietress was enchanted. The proprietor in our small hotel was not all that enchanted by the boys' preoccupation with the small open elevator in the hotel.

I also made a number of short trips to Europe to give seminars. My seminars are usually on some aspect of my efforts to understand properties of solids in terms of electronic structure, but for these I chose another topic, *The Meaning of the Question, How Long Does it Take?* It's an interesting question for which most physicists will guess the wrong answer when it comes up. It turns out that in quantum theory the answer depends directly on how you go about measuring it, different answers for different experiments. I think it went well. I've given talks on a similar quantum-mechanics question more recently called *Does an Inelastic Event Just Randomize the Phase?*

Before we went to Cambridge there had been a reception in Palo Alto for V. L. Ginzburg, a leading Russian theoretical physicist visiting Stanford. I expressed interest in visiting Russia and he enthusiastically said to come ahead. I indicated that I may need an invitation and he said: "You are invited!" I don't think that helped, but I began to look into the possibility. I located a Mr. Zonn in San Francisco who seemed to arrange visits to the Soviet Union so I went to him. He was a large man with a large voice who said: "If you want to go to France, don't see me. If you want to go to Germany, don't see me, etc. If you want to go to Russia, see me!" He could arrange tickets for me to Moscow and St. Petersburg (then Leningrad) and the required visas and hotel reservations, he seemed very supportive and enthusiastic. I had a grant which could cover the travel expenses. I finally arranged and took such a trip to

Moscow in December 1969, before we went to Cambridge, and again in September 1970 from Cambridge to Moscow, Kharkov, and St. Petersburg. I have a good record of the second trip, but only memories from the first, which are very much mixed with memories of the second. This account is based on my memory of the second with one or two thrown in from the first.

For this second trip I wrote to Genrich Krasko, who had translated my book *Pseudopotentials* into Russian, and would later translate other books of mine, but I had not heard back before I left Cambridge for Russia. I arrived in Moscow and after two hours of arriving at the airport I took a cab to the Metropole Hotel, next to Red Square. I walked around the area and had dinner at the hotel. As usual, it turned out that I had dinner with someone I could talk to, in this case a visiting Siberian.

The next morning I hired an interpreter to help me find my physicist friends. We first went to a bank where Krasko had told me on my first visit that royalties would be deposited, and withdrew 650 rubles, officially equal to $650, but on the black market more like $150. We found that there was no phone book and apparently no good way to get phone numbers. We *did* get Krasko's office address and were able to call the Academy of Sciences and the Physics Department at Moscow State University, but neither answered. Sometime in the afternoon we gave up and went to Gum, the major department store in Moscow, and I bought everything I could possibly want and carry away, a watch for Lucky, clocks for the boys and a soccer ball. The next day I took a cab to the Academy of Sciences, but had no success, and went on to Krasko's institute. After a considerable time they said that he wasn't in today, which I didn't believe, but turned out to be true. They didn't know his address nor phone number so I left my hotel room number, which they promised to give him. I returned to the hotel, and managed to arrange a call home, which went through and I could talk with Lucky. The next day after breakfast, to my surprise Krasko showed up. We went out for a walk, away from the listening ears in the hotel. I learned about his life and his reaction to the inflation that was taking place in Russia. There it didn't come from rising prices. If a pound of butter in a green wrapper is 60 kopeks, they introduce a new brand in a blue wrapper at 80 kopeks and you could buy either, though no one buys the 80 kopek brand. Pretty soon the 60 kopek brand is no longer available.

We had lunch in the Aregvi Georgian restaurant where I had eaten on my first visit, and went to his apartment. I remembered that on my first visit we had gone there and he had put a bottle of vodka, with a dried red pepper soaking in it, outside the window in the December weather. When we toasted with it I remarked that if any drink should be called a Molotov Cocktail, that was it. He had never heard that term; apparently the powers did not want the populous to know about that tool of revolutionaries. On this trip we just exchanged gifts. Krasko rode with me in the cab to the airport but en route the cab driver, presumably a KGB agent, got in a heated discussion with Krasko, stopped out in the country, examined Krasko's papers, and made him leave the cab. I was furious, but there was nothing to do.

I was flying to Kharkov hoping to see a Professor E. M. Kaner, whose name I knew from Azbel-Kaner Resonances which were used in Fermi surface studies. The flight to Kharkov was delayed two hours, and the passengers, many of them Africans returning to school, were drinking and making merry. Our propeller plane finally took off, and when we landed in Kharkov, we sat on the runway for over an hour before going to the terminal. The crowd remained jovial, I thought maybe the only way to deal with continual frustrations. When I finally found the baggage pickup there was someone starting to walk off with one of my bags, but I took it back. It was a passenger that I had talked with earlier, but he no longer spoke English. A girl named Natasha from the Russian Intourist service had been assigned to me and identified herself, and we left by car to the hotel. I don't think I paid for a guide, but she was probably associated with the KGB and there to keep track of me. I checked in at the hotel and met a group speaking English and drinking in the small lobby of the hotel on the third floor near my room. I joined the group, and brought my bottle of Applejack that I had been given by Krasko. They were mixed nationalities, but by late evening it was just two East Germans and me and it was easier in German than English and we talked until quite late. I sat with them again at breakfast.

Natasha showed up after breakfast and said it would be best to call the University ahead to make an appointment with Professor Kaner. She said that it would take two hours or so to arrange it, so she should take me on a tour of Kharkov, the second largest city in the Ukraine, with the Intourist car. I gave up on Kaner and enjoyed an extensive morning tour of the city. When we got back to the hotel, the manager informed me that there was no Dr. Kaner at the University of Kharkov, which didn't surprise me at all at this point. I decided to go to a Ukrainian restaurant for lunch and Natasha said it would be more "convenient" if she went also. The main dish was much like ravioli and delicious. I decided to go shopping again in the afternoon and Natasha was of great help. I bought a chess set, a pedometer, and a compass, and then went on to a historical museum. Back at the hotel I was astounded that the manager told me that Kaner had called and left a number. I called and he met me at the hotel. He said he had written to me in July inviting me to give a seminar, and indeed such a letter arrived some time later. He hadn't heard that I was there till then. One of his students joined us and we went to my room and exchanged gifts, and had toasts. We had dinner at the hotel, also with his wife, Irene. It was a pleasant evening but Mony Kaner was cautious about what he discussed, assuming someone was listening. Irene was excited about having just published her Russian translation of *Winnie the Pooh* and about how carefully she had chosen to translate "Piglet", to give the same feeling for Russians that Piglet did for English. Many years later the Kaners visited Stanford and we had dinner with them at the Kapitulniks. I sat next to Irene and at that time the conversation seemed to go better in German, which I enjoyed, even if it surprised me. Back in Russia, Mony's student agreed to send a wire to Igor Abarenkov in St. Petersburg telling him that I was coming.

The next morning I left at 7:00AM for the airport, and Natasha was there to accompany me. I told her that I had met the Kaners for dinner at the hotel, and she said yes she was there also. At the airport we learned that the flight was postponed and I didn't end up leaving till 4:00PM. Natasha stayed on and said she would be given an extra day off for the time she had spent. She planned to take it on her birthday, October 9, the same day as Lucky's. She said she didn't know how to play chess so I taught her on my new chess set. In her line of business, which I assume was being undercover agent, I've wondered if she was just humoring me.

The arrival at the St. Petersburg airport was as hectic as the others with huge throngs of people fighting to get their luggage. When I finally found the Intourist office they put me on a large beautiful bus with one other passenger, to my hotel, the Sovietskaya. The city was beautiful, wide streets with trees, another world from Kharkov, and Igor Abarenkov was at the desk on my floor. I had not met him before but he was magnificent, my age and a red mustache, looked like a Scotsman, and spoke perfect English. We had dinner on a balcony in the hotel and watched the dancers below. He had planned for my visit. The next day was Sunday and we took off to Petrokov to see the grand palace of the Tzars, built by Peter the Great. Back to St. Petersburg to meet Ija Ipatova, a physicist from the Academy. We had dinner at Restaurant Sadko with a delightful Russian orchestra, mandolins, accordion and a balalaika and a tour of the city at night.

Monday morning Igor picked me up, and had my entire visit scheduled. We met Ija at the Hermitage Museum, about which she was an expert. She said time was short but we could see the three most important items in the museum: first a *very* old cart, which she said was perhaps the oldest wheeled vehicles in existence. The second was a small Greek sphinx which had the original paint, which has disappeared from almost all ancient Greek statues. The third was a special icon painting, among an entire room full, which had a special significance that I can't remember. I gave my seminar on *How Long Does It Take?* at the University in the afternoon. On Tuesday we went to the Ioffe Institute for semiconductor physics where seven theoretical physicists each had a few minutes to describe his work, which we then discussed. I also gave a seminar on bonding in metals, semiconductors and insulators.

In the evening we went to the Leningrad Philharmonic to hear an East German string quartet concert. I was only mildly interested but during an intermission, walking around the hall with Igor, he said he was enjoying it much more than usual. He was letting his mind fly to other things and letting the music flow over him, instead of worrying as he often did. It had never occurred to me that a concert was a proper context for dreaming, but he was right! I enjoyed the last half immensely. We sat on the side wall and the faces in the audience were fascinating to watch. Wednesday morning Igor showed up with an Intourist car to take me to the airport. There were no problems at the airport, as with earlier stops, and I boarded the flight to Stockholm with many teenage Swedish tourists, carrying balalaikas. It was somehow a relief to pass through customs there and enter the western world before flying to England.

We stayed in Cambridge through Christmas and returned to California in January. We had arranged a stop in the island of St. Thomas on the way to Mexico, but when we got to the airport we found that we didn't have the needed visas for Mexico. We were rerouted to Miami, bypassing St. Thomas, to get the papers there and stayed overnight in a motel. When I woke up in the morning and opened the door, the Florida sunshine was so starkly brilliant that I immediately shut it. We hadn't seen bright sunlight since we'd left California. We proceeded to Mexico City where we had a large suite in the Hotel Reforma-Intercontinental. Before dinner Lucky and I left the boys to go down to the patio bar for a short drink. When we came back they had found a chair they called a throne and were playing royalty. We went on to Mazatlan where we had a seaside cottage in a resort with meals. Every morning the boys would set up our place for the day under a palm tree on the beach. The swimming and the food were good and it was a fine end to our sabbatical in England. We flew back home to Stanford and the life on campus.

Chapter VI

BOYS GROWING UP

The time around the 70's was turbulent, with student revolutions because of the Vietnam War. I had to think about the possibility of emigrating to Canada to remove the danger of draft into an inappropriate war for our sons. I had even picked the likely site if it was necessary, Simon Fraser University in Vancouver. Some of our sons later got active in antiwar protests in San Francisco and on one occasion I marched with them at such a protest against the Iraq war. Fortunately, our family has managed to just miss all the wars. I was too young for World War II, went to graduate school during the Korean War, and too old for the Vietnam and Iraq Wars. Our sons were all too young for the Vietnam War and after that the draft stopped. Chuck was not so lucky, going into the navy just as World War II ended and being drafted during the Korean war, but his duties were good; it was just lost time. I think the turmoil left me feeling I must do *something* for the world and I chose being a scoutmaster.

Rick had joined scouts before we left for Cambridge, in Troop 51, led by Dr. Herb Hultgren who lived also on San Francisco Court. I understand that it was the troop organized with the help of Jane Stanford in the early 1900's. Herb was quite the mountain man. He had several older boys in the troop and took them on strenuous overnight hikes or climbs, but I think Rick was able to keep up. When we went to Cambridge, Rick affiliated himself with a troop there and went to weekly meetings. The main activities seemed to be knots, fish and chips, pancake breakfasts, and a trip to London for a "scavenger hunt" on subways and a tour of Robert Baden-Powell's home. It was quite a contrast to Herb Hultgren's trips.

When we returned, Rick rejoined Troop 51 and John signed up also. That got me involved as a father going along on the trips. The first was snow camping up near Echo summit in the Sierras, my first time with a backpack and the first time on snowshoes. The next was easier, but my memory is that some of the scouts were organizing a snipe hunt and John wanted to go. I intensely dislike that kind of prank and discouraged him as much as I could, but I think he went along with it. Herb indicated that he would like to back off as scoutmaster and I agreed to step in. Herb continued for a while sharing the monthly trips, which helped me a lot since I was such a greenhorn. We had had only trips to the local scout camp when I was a scout.

Scouts at the top of Clouds Rest, Yosemite Park. John Storm, Bill and Bob in the foreground.

It was quite an undertaking for me to prepare programs for the weekly meetings at Escondido School. I don't remember it so well, but we talked about upcoming trips and sometimes there were movies, and sometimes we worked on merit badges. Once a year it was first-aid merit badge and Herb liked to do that. It was a big relief for me to have him do it and I rationalized that he was an M.D. and especially well qualified. Actually I thought he did it all wrong. He liked to talk about disastrous occurrences and what *he* would do, but of course *you* shouldn't do that. I thought the idea was for a scout to know what to do in an emergency and not panic.

Another part of the program was Camp Oljato, the boy-scout camp at Huntington Lake, two weeks every summer. I slept with our troop in a tent camp away from the center of camp. There was a family camp connected, and before Bill and Bob joined scouts, Lucky would take them and sleep in a tent right next to the lake. They also ate in the camp dining room. I could go off and join them before dinner and enjoy a martini before heading for the dining hall. It was a nice time to relax and refresh. Once Bill and Bob joined scouts, that extra feature was gone.

After a few years I was beginning to feel like I should pass on the reins. Twice I found a father who was interested and spent a year training him, but both disappeared at the end of their year. Finally David Coward, a neighbor with boys in the troop, took it on and was really in charge. He was actually a graduate of the Engineering Physics Department at Cornell, as I was, and was working at the Stanford Linear Accelerator Center, SLAC. Bill and Bob continued, as did John as Senior Patrol Leader, and I faded away. In the end all four boys were Eagle Scouts.

Ted Geballe initiated a joint Berkeley–Stanford physics dinner, initially at Jack's, a restaurant in San Francisco. The solid-state physicists in Berkeley reciprocated the next year, arranging one in a Chinese restaurant in Berkeley. This was a nice chance to meet up with many friends in the two departments and it continued for a number of years. The person responsible for organizing it for Berkeley rotated among Physics Department members, and at Stanford soon the switch was to me organizing it. Berkeley settled on the same Chinese restaurant in

Berkeley, but I took it upon myself to find a different ethnic restaurant, ordinarily in San Francisco, each time. We did Afghan, Turkish, Thai, Czech, Russian, Spanish, Basque, some in outlying towns, but never far.

One year my German friend, Hans Queisser, was visiting at Berkeley and I asked the Berkeley rep to ensure that Hans knew about it. This caused a surprising problem. Hans was visiting in the Materials Science Department at Berkeley, and only Physics Department people were invited to these affairs, which I hadn't known. They would need to confer within the Department to see if it was OK to make an exception, which it turned out it was. This situation hadn't come up at Stanford since we started with both Physics and Applied Physics, which at Stanford were like two regular Physics Departments. Besides, I had included appropriate people from Electrical Engineering and Materials Science, and the IBM and Xerox laboratories in our area. They all seemed to me like part of the Stanford solid-state-physics community, and also I was finding that we had smaller and smaller representations from Stanford's two physics departments, while the Berkeley contribution remained large. At one point, someone at Berkeley mentioned that they were missing some of their physics friends at Stanford and suggested restricting it to our two departments. He didn't understand our problem and I didn't explain.

When I thought it was time for me to stop organizing alternate years, I chose a young Applied Physics professor, Ian Fisher, to carry the torch. He seemed just the right sort of person, and further he was married to Yuri Suzuki, who was in the Materials Science Department at Berkeley which I thought might reduce the Berkeley rigidity. Ian and I jointly organized the next Stanford dinner and I bowed out. When it was our turn again, the dinner never got organized and there were no more Berkeley–Stanford dinners. Yuri left Berkeley for a position at Stanford and it was the end of the tradition. I sometimes wonder if I was holding it together alone, and the time had really come to let it end.

I received a call from Duane Wallace of Sandia Laboratories in Albuquerque, New Mexico, inviting me to visit the solid-state group for two weeks in the summer. It was exciting to go back to where I'd been in the summer years before, though not out in Coyote Canyon. I stayed in a motel owned and run by Chuck's brother-in-law, Joe Warner, on Central Avenue next to the Sandia Base. The group was interesting and congenial. Jim Schirber was head, and there were Al Swittendick, David Emin, and Galen Straub, all of whom became lifelong friends. We went out drinking some evenings, not always in moderation. That practice extended to March Physical Society meetings where we all went and Jim Schirber would seek out the appropriate watering hole for us to meet in the evenings in that city. Lucky would fly to Albuquerque for the weekend between my two weeks and we could explore northern New Mexico. These annual visits continued for a number of years.

During the school year I was invited to give seminars or colloquia at various colleges and universities, and went perhaps once a month. The host always paid the expenses and ordinarily nothing else. I would spend the day talking with faculty about their research, have lunch with several faculty, give my talk, and join a group for dinner. These were nice breaks in the routine, and not an overload.

We also had many visitors on Stanford campus, guests of various faculty members. Stanford was different from other places I visited in that the physicists did not get together as a group for lunch. Some went jogging, others ate by themselves, or in their own group. Many of these visitors had offices in Ginzton Laboratory, as I did, so I would round up the visitors to go to Tresidder Union for lunch. It proved to be an interesting and changing group. I remember one Englishman, Oliver Penrose, who is much more mathematically inclined than I am. Our discussions led me to a study of multi-atom interactions, more formal than my usual research, which I published.

Being in California meant that we were not far from the ocean, and we took advantage of it from the start. Initially it was day trips to beaches near Half Moon Bay, and occasionally we stayed at a house in Carmel, which we had learned about from our friends, the Jahsmans. However, soon we went to the Santa Cruz area and rented a house. One of the first was the Red Barn at Sunset Beach, perched on a hill overlooking the ocean. There was an upstairs dormer, overlooking the living room, as well as bedrooms and bath on the first floor. We could walk to the beach, down a very long set of stairs. The water was cold, and the waves could be large, but we could take short dips. We also learned about skimming on the thin layer of water left by a retreating breaker, with a circular plywood board. We had seen others do it and I reproduced such a skim board. Learning to skim inevitably led to some tumbles on the beach, but I was young enough to handle it, and the boys were old enough to learn. Eventually the Red Barn was rented out long term and we had to look elsewhere.

We soon found rentals at Rio del Mar, a couple of miles north of Sunset Beach. There were apartments in buildings on the road up from the beach, and we later rented houses directly on the beach. The apartments and houses were cheap enough that we could rent for a month, even if we were mostly there only on weekends. Just down the beach was the Cement Ship, resting on the sand, with a peer and walkway leading to it. It was an experimental venture at the end of World War I, and never did service. The waves at Rio del Mar were not as large as at Sunset Beach and it was friendlier for swimming. We rented many houses up and down the beach, but ended up with a regular, 270 Beach Drive, a two-story that could sleep eight people in beds, and more on the floor. We generally went once in the fall and once in the spring. We had many guests, mostly friends of the boys, and beach fires at night. As the years went by, we sometimes rented the house next door, 272 Beach Drive, a one-story house that could accommodate almost as many people. The cost of the rentals was continually rising all of this time and in the end we stayed only in 272, and usually three or four nights.

We also took a trip to the mountains each winter for skiing. The boys had been too young for real skiing in Schenectady, but I went whenever I had a chance there and Lucky took some lessons. In California, we could rent a ski house in North Lake Tahoe and as each boy got old enough, he got skis and could take some lessons and start on the slopes. Our dog Pokey seemed to enjoy the trips also, but not as much as the beach. When we drove to the beach he could smell it from several miles away and got excited. One by one, the boys could ski with me, and I'd stop and wait for them. Of course before long, they would have to stop and wait for me.

I don't know how we got interested in Estes rockets, maybe from Duane Wallace at Sandia Labs, but for a number of years we did regular launches. There was the fun of making and painting the rockets themselves, and then with battery connected to the wire fuse for the little rocket engine we could launch the rocket, going 100 or 200 feet up and parachuting down. In the backyard, or in the field, there were times when it landed on a roof or we lost sight of it, but if we launched them in the bed of Lake Lagunita at Stanford when it was dry, we never lost them. I never see them nor hear them now so they must have gone out of style. The boys also, particularly Rick, made hot-air balloons of tissue paper, with a lamp below to make hot air and launched them. The fire hazard made me nervous but as far as I know there was never a problem.

We also joined the sailing club in Lake Lagunita. That was particularly satisfying when the boys were quite young; Rick was at the tiller, John held the main sheet, and Bill and Bob each had a jib sheet. The lake is so small that we had to come about very often and all the sheets needed to be redone. It was a very busy time.

Part of the turbulence of the 70's was of course drugs, which were completely absent during my own growing up. We were aware of the problem, but probably not of the extent to which our sons were involved. There were some scouts who seemed more druggy than others. I think it was at the beginning of my second year as scoutmaster when the new scouts showed up, it looked like a very motley group. I thought maybe the new group always looked that way, but they don't. It was just that one year and the group didn't last long. The next year the new group was great, and it included Bill and Bob. Once I picked up a stash that some scout ahead of me had dropped on the trail. I have to admit I had to try it out, and did smoke pot on a few other occasions with friends, or in later years with our boys. Once at a scout meeting two mothers were at the door and asked if I knew their sons, which I didn't. One mother said: "I thought so, but my son says every Wednesday evening that he's off to the scout meeting." Our family got through pretty well. There were a couple of incidents in high school involving the police, but no long-term problems.

I mentioned John Ingelsfield's trip to East Germany when we were in Cambridge, and my inquiries about going. I was indeed invited by Paul Ziesche for I think the third such conference, called a "School", including a number of East German college students. It was sponsored by the Dresden Technische Hochschule but held in a palace (they called it a castle) in the little town of Gaussig near Bautzen. In those days getting to Dresden was not simple. In the first place I arranged to pay for the trip with an Army contract I had, and in the middle of the arrangements, my army contractor told me that they couldn't pay for a trip to a country behind the Iron Curtain. I had to find another conference at that time that they could pay for and I would pay my expenses beyond that. It seemed an impossible situation, but I did find one in Köln and presented a paper there too. It turned out to be a somewhat interesting conference and I had friends attending. Then I flew to Berlin and went by tram to Friedrichstrasse Station in East Berlin and then by train to Dresden, where I had booked a hotel.

I was in touch with Ziesche and he had arranged for a younger physicist, Helmut Eschrig, to look after me. He and his wife, Rose Marie, had a young son. She had lived in Dresden and had survived the overwhelming bomb attack by America during World War II. She had learned in school that we had destroyed Dresden in order to slow the Russians down so we could get to Berlin first. I told her I felt sure that that was not true but I would try to find out the reason when I returned. I did explore the question with historians at Stanford and in the end I concluded that there had been a meeting, led by Dwight Eisenhower, to consider the options. I expect that, as at a Department meeting at Stanford, people expressed a wide range of opinions, possibly including the one Rose Marie had learned. Then when it got late, they took a vote and, as at a meeting at Stanford, no one could say why it came out as it did. I have remained friends with the Eschrigs since then and even spoke at a memorial service for Helmut in Dresden when he died a few years back. Rose Marie and I still exchange notes at Christmas.

I was there before the school started and was asked to give a seminar at the Technische Hochschule. I had volunteered to give my talk in Gaussig in German, but Ziesche said that was not possible; there were attendees from many countries and all talks at the conference were in English. However, I could give my seminar at the Technische Hochschule in German. I *did* that, which was even a little rash, I think it was a good idea. I discovered that I didn't know technical words in German, just Kneipe Deutsch, language you use at a bar, so I proceeded in German and used English for all technical terms. It slowed me down enough that I think they understood me much better than if it had been in English.

The conferees took a bus to Gaussig and I had the Graf's bedroom (the bedroom of the Duke who had owned the castle), shared with a Russian, Mauny Bravmann. He said it was his first trip *abroad*. I was surprised that a Russian would consider East Germany abroad. Our meals and the talks were all in the castle, and we got some exercise in the front yard. I had brought a Frisbee to the meeting and it was brand new to all. When we left the conference I gave it to Norbert Winkler, a young boy who hung around the lawn to see these out-of-towners.

The little town of Gaussig didn't have a gas station, but it had two Kneipes and a group of us, including Helmut Eschrig, got very familiar with one that had four tables, as I remember. It was the custom when one entered the bar to knock on each table, and each local would knock on our table as he came in. The same locals came every night and we knew them all. One brought a small dog that he said smoked. He demonstrated by holding the dog's mug shut on the cigarette and closing the nostrils. The dog tolerated it but I don't think he enjoyed it. I even met Herr Pötig, the shoemaker, when I went to his shop for a shoe repair. He sent me a postcard that turned out to be remarkable. I was given a woodblock print of Dresden by Rose Marie's father, which now hangs in my office; if you place Pötig's post card on top of the Frauenkirche, in the print it exactly matches.

The attendees at the school included a number of people from other East-European countries, and a few Scandinavians. At some point during the week someone said that I was the only person there who did not speak the native language of the conference, broken English. I objected; there was one Swede there who spoke much finer English than I did. Bravmann and I got on well, and shared a hotel room in Dresden for one night after the school. In East Germany twin beds come exactly next to each other. In the morning he said "Oh my God, I've been in bed with an American Capitalist!" He made a request that I couldn't see how to grant. He wanted a pair of eyeglasses exactly like Bob Schrieffer's glasses.

After the meeting I continued by train down to Prague, to see Czechoslovakia. On the train the conductor examined my passport and asked where was my exit-visa from Germany. I had no idea because others had made all of my arrangements and he said I would have to disembark at the border unless he found it. I couldn't believe it and had another beer with the people I'd met on the train. The conductor finally showed up again, said he had found the exit-visa, and returned my passport. When we paused at the border, there was nothing there but a stand with a guard holding a submachine gun. That really gave me a chill.

I stayed over in Prague and the next morning was May Day. It didn't mean much to me but it's a huge communist holiday and Wenceslas Square was filled with people holding little flags. I watched for a while and then had a beer in a pub close by. I sat next to an old Czech engineer who was interested to chat. The only language we had in common was German and I greatly enjoyed the conversation. I gave him my ball-point pen, that had Stanford University written on it. He later wrote me a post card to Walt Harrison, Stanford University, USA and it got there! The stop in Prague ended my first visit to Gaussig. I returned again to the school two years later, and I heard that Ziesche had told someone that I attended every other year.

Back in Stanford and in the 1970's, the boys were in high school. Rick had friends we knew and I remember chaperoning dances which he attended, but I have little memory of his social life. John's I remember better, including Ann Creger whom he dated for some time and they even took a trip to Hawaii together. He also invited Liz Marks to join us for a while at Fallen Leaf Lake. We were much more involved with Bill and Bob's life. They had a sizable group, including Debbie Lehman, Julia Eisenthal, Serena Clayton, and Cathy Coward, daughter of my successor as scoutmaster. The parents took turns providing them rides to various events, and I remember stopping the car at some suitably safe place to let them all run around the car and reenter. They called it a Chinese fire

Bob, Bill, John, and Rick at Fallen Leaf Lake, 1980.

drill. Jenny Krumboltz (now Somerville) was a special friend of Bill's whom Bill and we have continued to be in touch with ever since. She even helped with the design for the cover of this book.

There was also Bob's friend Jeff Melchor, whose father, Jack, owned a ranch south of San Jose. Once the Melchors invited the entire group to stay over at the ranch for a weekend. Jack wanted someone to share the chaperoning and I agreed. It turned out to be a most interesting weekend for me with Jack. As a boy he had come west from the poverty-stricken Okies and swore never to be poor again. Indeed he became quite rich doing business connected with the growth of silicon valley. He also had a keen interest in physics and we spent the weekend in wide-ranging discussions. I haven't seen him since but I remember bringing his name up again somewhat later. I read and was very much taken by Kurt Vonnegut's novel, *Cat's Cradle*. I somehow felt attached to Vonnegut. He had been a DU at Cornell somewhat before me, and he had worked at General Electric in Schenectady, also before me. One of his short stories was based on the Alplaus firehouse, which we passed every day on the way to work at the General Electric Research Lab. I even used his stories as a theme in a series of talks I gave at William and Mary College, at the invitation Arden Sher. When a newer novel came out from Vonnegut which I thought was not nearly in the same class as *Cat's Cradle*, I thought he needed inspiration. I wrote him a letter suggesting that the growth of silicon valley might be a basis for one of his novels. I invited him to visit and have lunch with the some of the principals, Jack Melchor, Bill Shockley who was coinventor of the transistor and a friend of ours, and some others. It was silly, and of course he never answered my letter.

Then some years later, Lucky and I were walking down the corridor in the Claremont Hotel, uphill from Berkeley, and Lucky said: "Wasn't that Kurt Vonnegut who just walked by?" It was and he stopped at a desk to inquire about something. I approached him and introduced myself, saying that we both had been in the DU house at Cornell. He said something like "Oh?" and stepped out on the balcony for a cigarette. I doggedly stepped out also and told him about the letter I had written him. He said: "That's interesting. Did I answer?" I said no, and after a puff he said: "I am the most famous person to have survived the Dresden fire." I thought, but didn't say: "The hell with you!" and departed. Not a good end to the story.

Joshua Slocum in the Spray.

Another author who seemed to show up often in our lives was Joshua Slocum, the author of *Sailing Alone Around the World*, a book that everyone who sails seems to know. He kept his sailboat, the *Spray*, in Manchester-by-the-Sea, Massachusetts, at the same time that my grandfather, Walter Ashley, was a minister there. They were acquainted and Slocum sometimes came to dinner. Gertrude was then in her early teens and very much taken with him. She left the copies of her diary from that time in the autographed copy of his book, which we now have. I read the book at about the same age she had been, when I was sailing in Lake Erie with Harry Kerr. Harry also had a copy and when he died, his wife passed it on to me, so we had two copies and most of our boys also read it. When I retired we held a large retirement party, much like our Oktoberfests, and Barbara Kerr was there. I asked her if she'd like Harry's copy and never saw anyone look happier! She clutched it all evening.

I had had an Indian physicist visiting with me for a while at Stanford, a Professor Gupta from New Delhi, and I had a yen to visit India. There was an International Research Exchange Program that would fund such visits and I applied. It took a year or so to hear, but it came through and I planned a trip. It was only easy to go during our summer recess, which isn't the best time to visit India, but that's what worked. The trip was centered on the Indian Institute of Technology, the IIT Delhi, where Gupta was. India is about halfway around so it was natural to go around the world, and make stops with physicists along the way.

The first stop was Tokyo and on the plane I got talking to a Japanese man who was very anxious to show me Tokyo. He said the greatest eating is in taverns which are recognizable by a row of orange lanterns over the door, and they all have knotty pine tables and benches. He steered me to my hotel and took me to such a tavern and I never even could find out how much anything cost. I left the next morning for China, not sure that I had really seen Tokyo.

Shang Yuan was still at Stanford and had helped arrange a visit with a Professor Xi in Fudan University in Shanghai, where I gave a talk. This was early enough that few western visitors were coming so I got lots of attention. I also had time to explore Shanghai. At first I was puzzled that though everyone I could see on the streets was Chinese, nobody seemed to notice that I was there. Then sometime when I was standing looking at something, a girl who was studying English approached me and wanted to try out her English. As soon as we were talking a large crowd gathered around. I wasn't invisible after all. I realize now that my mode of travelling was rather unusual. Others I have talked to were usually with a tour group or had a guide of some kind. I wandered around the alleys where men were sitting in front of their abodes cutting up food for dinner. I could eat in the hotel but could also walk to someplace and sign-language my way through to dinner.

From Shanghai, I flew on to New Delhi. I happened to sit next to an older Chinese engineer on the plane. He didn't speak English, but he *did* speak German, having worked in Berlin before World War II. We had a nice chat on the way. We arrived in the evening and I asked a cab to take me to the Indian Institute of Technology, the IIT. He acted like he knew where we were going but it was a strange trip, asking passers-by for directions, and cattle wandering freely through the streets.

We did get there and I had a nice room, with open windows and door to give some relief from the heat. I was struck by a couple of decorations on the walls, live lizards waiting for flies, but one doesn't notice them in India. I talked with a number of students, and faculty, about their work. It was a bit run-of-the-mill. I had meals in the institute's dining room, which seemed a little like a colonial dining room, waited on by Indians in white coats. They didn't serve beer but I managed to buy some outside and they kept it cold for me. When I would open the refrigerator door to fetch a bottle I'd see cockroaches skitter away inside, but I guess one doesn't notice that in India.

I could also go out in the evening and walk around the city. One evening I walked past a dark area and noticed a pile of pots next to the sidewalk. Suddenly a little girl emerged from behind the pots and walked to the curb to urinate, and then went back presumably to where she was sleeping. There was poverty all around but that was the first time that it seemed overwhelming to me.

I also travelled to a number of IIT's around the country, which gave me a chance to see India. I remember particularly Agra. I went by train and made the mistake of buying lunch from a vendor on the train, so I was sick part of the time in Agra, but not all. When I started my talk at Agra College I saw the glazed eyes of the attendees and realized that they were not even going to try to understand what I was saying, just go through the motions of a seminar. I cut it rather short, and I expect that they were relieved. At the end of my talk someone got up to ask a question, which he read: "J. C. Phillips, in the USA, said ...", something about my work. I responded and everyone was happy.

I saw the Taj Mahal, which was certainly impressive and in those days completely white. Also in Agra they were building a new temple in white marble, with pieces formed in the old way, by hand. It wasn't expected to be finished for a couple of hundred years. On another trip I took a train to Jaipur to give a talk, and had the customary elephant ride in Jaipur. I also went to Varanasi and saw the Ganges River, with the steps leading into the water. No funeral pyres burning that day, but lots of people bathing. I wasn't tempted. Near there I bought my copper water jug of the shape Indian women carry on their heads to bring water home from the well at the center of their village. They are mostly made of pottery, but more prosperous families can have brass. Only the elite can afford copper. It still sits in our living room. It was my largest purchase in India, I think around $25.

My last visit was at the IIT in Bombay, now Mumbai. I was to fly from Bombay to London. Earlier I had planned stops in Tehran, Iran, and someplace in East Africa on the way but there was political instability in those areas so I cut them out. I went to the Bombay Airport at the appropriate time and there was fire-engine activity there. It turned out to be serious, with a couple of people killed and they closed the airport indefinitely. I returned to the IIT and tried to get a train out but so many others were trying that it wasn't possible.

I was able to continue staying at the IIT, and took the afternoon to try and swim in the Indian Ocean. I took a bus to a beach near the city and that was interesting. They packed unbelievable numbers of people in the bus with some hanging on the outside, but I was surprised there was no smell of people and sweat in the bus. I figured the people in India must bathe a lot and it wasn't like a bus in Paris. At the beach the water was very shallow for a long way out. I managed to get wet, but it wasn't very successful. A couple of young kids followed me wondering what I was up to.

I learned that there were two planes at the airport and they would fly out the next day, one to New Delhi and one to Madras. I signed up for both and went with my suitcase the next day to the airport, though it seemed very unlikely that I would get on either. I was astounded to hear my name called and I was off to New Delhi and then London. I stayed overnight in London and, surprisingly, the first order of business was to go to Wardour Street in Soho, where I knew there were many Indian restaurants, for a spicy lamb curry. I had been served lamb curry many times in India, always with an apology for Indian food being so spicy, but always the curry was much too bland. It was much better in London. From London I flew to San Francisco and home, completing the trip around the globe.

When it came time for Rick to go to college, he applied to engineering at the University of Colorado in Boulder and was accepted. We realized that he hadn't taken the calculus courses that were expected there and we worked on it at home, where he had moved into the Annex apartment in our house. He picked it up quickly, but I don't think it was nearly as good as a more leisurely course. He went off in the fall and everything seemed to be going well. He was dating a twin coed. I visited on the way to a seminar in Colorado Springs and we had dinner together. However, he was not as happy as I thought and was recruited by Scientology. When he was home for Christmas he announced that he was not returning to school. He did return to Boulder but to work with Scientology. He asked for, and got, the funds that had been given to him by my Uncle George for his education. His education just turned out to be different from what George and I had imagined. He later moved to Los Angeles, associated with the Scientology organization there.

When Rick left for college, John moved into the Annex. When he was ready for college, he did apply to Cornell, but was more interested in Colorado College in Colorado Springs. He was admitted and accepted. That was particularly interesting to me since my father grew up in Colorado Springs and both my father and mother had attended Colorado College.

My father had hoped to take Chuck and me west to Colorado after World War II restrictions on travel ended. However, he was unable to get away and sent us on our own in 1946 or '47. We took the train out (coach) and went to look up my Uncle Frank Harrison, whom we had never met, in Colorado Springs. We found his office and he was sitting in his shirt sleeves at a desk near the door, with a man in a suit — clearly his boss — sitting in a back office. We got acquainted and pulled up chairs to talk about where we might do an overnight hike in the Rockies. Frank had a suggestion but wasn't sure and signaled the man in back, who came running. "Yes, Mr. Harrison?" We had misjudged who was boss. Frank did drive us up the canyon above Colorado Springs from which we hiked up finally to a meadow and slept in a deserted and ramshackle house. Nobody else was around but it was a natural place to go from Colorado Springs and presumably the meadow my parents had hiked to in their youth. On this trip to Colorado we also visited my Aunt Kate and Uncle Arthur in Denver, who took us to Estes Park, but Colorado Springs was the best part of the trip.

John went off to Colorado in the fall and into a college dormitory. Lucky visited once during the first year. Colorado College had an unusual schedule, with one course exclusively for several weeks and then a new course. It seems to work as well as the usual scheme of several courses at once.

John had many friends and enjoyed being in Colorado Springs. He had a friend Stuart who rode in freight cars on freight trains, and John joined him, one time coming to California that way. It made me nervous when they arrived with sooty faces, and I was very uncomfortable dropping Stuart and John off at the rail yards in Oakland for the return trip.

John took all the art courses they had at Colorado College and after two years felt that he should move to The San Francisco Art Institute, which he did, and he finished there. His specialty was silk-screen printing. That had begun much earlier with our printing of silk-screen Christmas cards, which Lucky and I had done from the beginning of our marriage. John remained in San Francisco, initially as a silk-screen printer.

After John left there were just the four of us at the dinner table; we called it the regular four. When their time came, Bill applied to the University of Colorado and Stanford. Bill took a trip to Colorado to visit, but found the atmosphere depressing. Stanford was definitely not and when he was accepted by Stanford he said yes. It was a contrast to the others who had gone away to college, but he moved to a freshman dormitory, Donner, and we didn't see much of him till Thanksgiving. He seemed to change dorms every year or so and graduated on schedule.

Bob applied to a small college in the northwest and to the University of California at Davis, but he had many friends going to UC Davis and chose that. He started at Davis, but left after a couple of years. Struggling with alcoholism and drug addiction during this time, he lived at home for a while working at Mt. Eden Nursery in Mountain View, then bought a vintage Volkswagen van and moved to San Diego. He moved back to the Bay Area after a few years, and got sober, partly with the help of the Redwood Center in Woodside. The only way we could see him at the Center was by going first to an Al Anon meeting at the Center. We quite enjoyed the meetings, as well as seeing Bob. It was a very articulate and interesting group of people who attended.

Our earlier experience with Al Anon had not been so good. It was at the Lutheran church not so far from our house and Lucky and I walked over there after drinks and dinner one evening. It turned out to be a meeting for adults with alcoholic parents (or ACA: Adult Children of Alcoholics), not a meeting for parents that have an adult child who is an alcoholic, as we had thought. We had gone there to support Bob, but as it turned out we were viewed a little as the enemy. Bob went to work for Bank of America in San Francisco and eventually completed his Associates Degree at City College of San Francisco.

During this period the question of sexual orientation arose. Since all four of our sons had dated girls throughout junior high and high school, we had every reason to assume they were all straight. There may have been early signs, but during this time we found out for sure that three out of four — John, Bill and Bob — were all gay! I think the first occasion was when we stopped at John's apartment in San Francisco to pick him up and he was still in bed with his friend Sam. We didn't think that much about it but he felt it made clear that he was gay. Later he met Lucky for dinner in the City (San Francisco) and explicitly announced it.

In the mean time Bob had called me from Davis. He said he needed to talk to me and could we meet in the City. I didn't see why we couldn't talk on the phone, but he insisted and we did meet for dinner. He announced at that dinner that not only was he gay, but he had joined Scientology. He told me that he didn't want to be gay and that he thought Scientology could "cure" him. Apparently the folks at Scientology thought that surely a father would be willing to throw money at the prospect of salvaging a son gone, in the Scientology view, astray. But I thought they were wrong. Our memory is that I told him that there was probably nothing he could do about his sexual orientation, that it was genetic, and that he might be better off just accepting it.

Bill told us he was gay after he had finished Stanford, and was about to go to Portland, Oregon, for a short-time job. As they each came out and throughout these years, our main concern for our three gay sons was that their lives would be harder for them than if they were straight. We supported and accepted them as gay from the beginning, just as we have supported and accepted our son Rick as straight from the beginning. Through the younger three we have opened our family and our hearts to an ever-expanding extended and modern family. Of course each time a son came out it meant something different about the future than we had anticipated, but I think that future turned out to be very, very good.

We had a visit from my uncle, Richard Knight Harrison (II) [the same name as Rick (III) and my grandfather (I)] and his wife Jerry. Jerry also had a nephew in the area, Greg Allen who had a friend Phyllis (Foo), so we all went to dinner together. Greg and I had an interesting relationship, with his aunt married to my uncle, far from a blood relationship. Once at a wedding in Toledo a group of us, including our Bob, were talking and I realized and pointed out that this nephew of Carolyn had the same relationship to Bob that I had to Greg Allen, very remote.

Greg and Foo became long-term friends. After my Uncle Richard died, Jerry came through to visit Greg and me and I had the chance to take her out for lunch, the last time that I saw her. Greg had a Cajun friend, Beau, and his wife Cathy, and another couple whom we also saw regularly. Greg and Foo came for years to our New Year's Eve parties and Foo was a conspicuous addition, once with such a flamboyant jacket that everyone had to try it on. On one occasion, Greg and I launched a new project, to write a new chapter for the bible, and to put in it all the things that *should* have been in the original. We had in mind universal truths, like "Shit happens!" We never put anything on paper, but I thought we ought to have a location in the bible in case, and spontaneously came up with *II Corinthians 13*. When I later checked, I found that the original *II Corinthians* ended with Chapter 12. It seemed that someone was on our side.

Greg through his work had come to know a number of local Portuguese people who held a cioppino fest annually in South San Francisco. It began with a no-host bar and then a dinner at very long tables with unlimited Dungeness crab cioppino and red wine. We went every year for many years. At some point Foo left Greg and went off with Frank Cooper, and we ended up seeing more of Frank and Foo than of Greg. Then Foo left Frank and when she was diagnosed with breast cancer Greg took her back till she died. Later Greg moved with a new lady into the Sierra foothills. The last time we saw Greg he came to our dwindling New-Year's-Eve party on a walker. We were so short on men to put on *The Three Billy Goats Gruff* that he and his walker did the bridge by themselves. Greg has since died. All those connections through my remote cousin Greg have been quite something!

After Foo left him, Frank met Lila, an ex-ballet dancer from Yugoslavia. We saw a great deal of them and then they decided to get married. They were short on funds and we offered our house for the wedding and reception. I also agreed to perform the ceremony and applied to be a minister in the Universal Life Church, as Bill and Bob had done for other weddings. All you needed to do was sign up on the web, and Bill and Bob each paid $25. When I did it, it was free. Many years later we were at a wedding for Bob and Phyllis White's son and I found that the minister was also Universal Life Church, but she was making a living at it. After the service I commented to her that we were of the same faith, but she didn't seem very pleased.

Frank and Lila did all the refreshments for their wedding and when Frank was bringing things into the house he heard bagpipes playing down the street. He followed the sound and found a teenage boy playing bagpipes in his backyard. Frank hired him for the wedding, I expect it was the first gig for this young man. Frank completed the picture by arriving in a kilt, and the wedding party marched into the backyard where the wedding was held following the bag pipes. Frank worked for some years as a technician at SLAC. The Coopers had an apartment in Mountain View and we saw them often for dinner. Eventually he retired and they moved to the Sierra Foothills and we see them when they are back in town.

We had decided it was best not to go away on sabbatical again while the boys were living at home, but when the time came for another sabbatical before they left, around 1975, I took a quarter or two off from teaching. It gave me a chance to travel through Europe for a month or so while Lucky stayed at Stanford with the boys, like a mini-sabbatical. I flew to London, visited Imperial College there, and of course Cambridge. I went on to the University of Paris, to visit Robert Pick who had made the extensive visit with us at General Electric, and then on to Stuttgart. There I visited Alfred Seeger in the Max-Planck Institute which was then in downtown Stuttgart. Alfred was well known as someone who didn't believe in pseudopotentials, but we got on fine. I enjoyed getting acquainted with downtown Stuttgart, including a downtown beer hall on the main street, Königstrasse, where I walked in late as everyone was leaning side to side singing and drinking beer out of large steins.

I had a free weekend after that and decided to rent a car and drive down to Salzburg and to the salt mines in Austria. I wandered around Salzburg and noticed a little sign on one house "Mozarts Geborenhaus". I felt a satisfying feeling of history being there where Mozart was born. I went to a nightclub after dinner where a comedian was making unmerciful fun of a patron, to the uproarious laughs of others. I decided I didn't like Austrians very much.

The next day I started south and saw three girls hitchhiking, so I stopped to pick them up. We had a little trouble at the outset communicating till one asked if I spoke English. It turned out she was a Stanford student studying in Europe, with another student from America, and a German student. They wanted to go to Hitler's Eagle's Nest at Berchtesgaden. They weren't interested in the salt mines and I decided to go to Berchtesgaden. When we got there we found that it was too early in the season for the Eagle's Nest. The roads were still closed for the winter. We had a nice lunch there and I drove them back to Salzburg. It occurred to me later that I had taken three unmarried women across an international border twice, and I suspected that the rather scruffy German girl was carrying pot, but nothing happened. That June, when I marched in the commencement procession at Stanford, I was hailed by my Berchtesgaden companion among the crowd of graduates.

From Munich I went on to Vienna for a visit to a university there, and then on to Budapest to visit Belos Vasvari, the Hungarian physicist who had been visiting Cambridge while we were there. I gave a talk in his institute, and the next day he took me to a studio where I was interviewed for the Hungarian television network. They asked me questions about the future of solid-state physics and translated the replies into Hungarian. I thought about how tedious these programs in such a communist country must be.

One night in Buda, after dinner, I went for a beer in a tavern near my hotel, and sat at an empty table. When it got crowded, a group asked to join me at my large table. They were a jolly group celebrating a birthday, and were interested to have an American with them. Again the only common language was German and we did fine till late in the evening. I don't recall the path back from Budapest to Stanford, but these were the high spots of a lively substitute for a sabbatical in the midst of our sons' growing up.

Chapter VII

SABBATICALS AND GERMANY

There is an old story concerning an argument between a priest, a minister and a rabbi about when life begins. The priest said that it begins with conception. The preacher said no, it begins at birth. The rabbi said no, neither of you are right, life begins when the children leave home and the dog dies. It was different for us; John, Bill and Bob remained within easy reach and we always stayed close, and we were always in touch with Rick. We waited a year to see that Bill and Bob's lives away from home had stabilized. Then for the second year we planned a sabbatical in Stuttgart.

I contacted Manuel Cardona, an old friend who had moved to the Max-Planck Institute in Stuttgart as Professor. Manuel started the application for a Humboldt Senior Scientist Prize for me, perhaps called "Prize" so it wouldn't be taxable. A little later he told me that it was better to have me associated with Ole Krogh Andersen, a solid-state theorist in the Institute whom I knew slightly. Manuel wanted to invite Jan Taucs, a physicist from Prague, and he could only do one. That was fine with me and turned out to be *very* good. We arranged to go after the fall quarter of 1982.

One of Ole's colleagues, Niels Christiansen, arranged an apartment for us on Gallenklingenstrasse in Botnang, a basement apartment but opening up on a backyard, with a view out across the valley to Ruble

Hill, where the remains of old Stuttgart were carted after World War II. Our landlord was Dr. Heid, a formal Schwäbish gentleman, who called me Professor Harrison and I called him Dr. Heid. Next door was Hildegard Wieck whom Lucky soon befriended. She was an artist and we have two of her prints on our wall in the hall. Her husband was the first violinist in Stuttgart symphony.

As part of our von Humboldt award, we had access to a new BMW at a very low price, which we picked up in Munich on our way to Stuttgart. I was later distressed to hear that we could only keep it for 1000 miles, but when I asked about that they said I just needed to turn it in in Munich and they would give us a new one. When we did turn it in, the clerk examined the car and said we needed to fill out an accident report. What accident? He found a tiny chip out of the paint on a front fender, presumably from a pebble that had been kicked up by a passing truck. I had to pick a date and time for when this accident occurred, but fortunately there was no other problem with it.

My office was with Ole's very congenial group in the Institute in Büsnau, a mile or two from our Botnang apartment. We in Ole's group had lunch together every day in the Institute's cafeteria, and outside when the weather permitted. Every afternoon there was coffee in Manuel's group, two floors up. We had regular seminars and I of course gave one at some point. I started a project with Ole, and before very long, I had a manuscript on our joint work with him. That turned out to be good fortune because Ole was a little pushy about getting people to work on something that interested him, but it took a long time for him to deal with my manuscript and that took all the pressure off of me.

I presented many talks at neighboring institutions, and Lucky and I made several jaunts in our BMW. One of particular interest was to Zurich, Switzerland, where a fest was held for Werner Känzig, our friend from Illinois and Schenectady. Werner had focused his work on some little-known materials, the monoxides of alkali metals. He had a series of students doing very interesting projects and publishing their results in German in *Acta Helvetica*, which almost no one reads, perhaps. He planned to write a book in English summarizing all of this interesting work. I tried to persuade him that this was a very bad strategy: once all this work was completed there would be nothing else to do and then nobody would be interested. It was to no avail. I think I was right, he never wrote the book and I'm one of the very few who know of these interesting materials.

We had many visitors during our stay. Our being there was the spark that ignited a European trip for several friends. Bill and Sally Marriner, whom we knew from Fallen Leaf Lake, came with their daughter, Beth. They stayed in a hotel nearby and then rented a car to tour southern Germany, and we met them in Garmisch-Partenkirchen in Bavaria. We did the baths, and Sally was particularly struck with the somewhat authoritarian attendants: "You will now go to the dressing room and change into the prescribed suit!" It reminded her of the World War II extermination camps. We were also visited by Susan Rice and Sally Chandler who had roomed together in our annex but were now at two different Stanford campuses in Europe. They stayed with us in our apartment and came to lunch at the Max-Planck Institute. I was quite impressed that Susan seemed to converse comfortably in Spanish with Felix Yndurain, a Spaniard in the group.

We took many trips around the area. The Black Forest lay just south of Stuttgart and was a colorful start. Our friend Jan Taucs, the Czech Humboldt visitor in Stuttgart, went there for Fasching, Mardi Gras in French, though we didn't go that particular day. There was a big crowd in the village for the parade and Jan was poking around to find a place where he could see the parade. A policeman stopped him and Jan took out his "Ausweiss". We had been joking about this identity card for visitors, indicating our country of origin, which we needed to obtain at the police department. It seemed so useless and so German. The policeman was impressed and thought Jan must be very important. He ushered him up to sit on the podium with the mayor to watch the parade. That is when we found out the value of an Ausweiss! We did travel to Tübingen, also south of Stuttgart and an ancient university town. Lucky bought a glass oil lamp there which we still have. In another small village near Stuttgart, they still had a sign indicating "Judengasse", the alley in which the Jews had been required to live. We also visited Heidelberg, the charming old university town north of Stuttgart.

When the time came for the regular school in Gaussig, East Germany, we planned to both go, as did a young Swede, Ule Gunnarsen, in our group and his American wife. When we got to Dresden the ladies were not initially allowed to come with us to Gaussig, but they got on fine in Dresden. Soon the organization relented and they were allowed to join us there, staying with us in the castle. In the evening we returned to the Kneipe, and indeed Norbert Winkler, who was the young boy to whom I had given the Frisbee, was there and joined us, now grown up and a mason by trade. He also seemed to have become a heavy drinker and was not well thought of in the village. The Kneipe owner Andreas Kusch, was pleased to meet the ladies and, with his wife, took them on a tour of the neighboring city of Bautzen one day.

After the meeting, back in Dresden, Lucky ventured out to visit the brother of a friend of Lucky's in Stuttgart who had asked her to look him up. On the way there, she passed a store with a big line and thought they must have something desirable, so she joined the line. It turned out to be rolls of aluminum foil and she bought one. This brother cried with joy when she presented it to him, apparently a rare treasure. We had driven our BMW there and were *very* carefully checked when we exited the German Democratic Republic. I think they were making sure we didn't sneak someone out with us.

At the end of our visit, we decided to hold a party for our Stuttgart friends. We thought the backyard by our apartment would be a good place but by then I had learned to be cautious about it, so I asked Dr. Heid permission. Wow, was I right! He said he would have to consult with the neighbors. He did and permission was granted. We bought food for a barbeque and I reserved a small keg at a local store. I arranged to pick it up on the afternoon of the party so we would have it very cold, but when I did pick it up, it was room temperature! Cooling seemed not to be a service they provide in Germany. We asked and there apparently was nowhere to buy a big chunk of ice. We kept it in the bathtub with whatever ice cubes we could produce. Everyone seemed to have a good time. Even Dr. Heid came down for it and took a small group up to his living room for schnaps (German brandy).

We left Stuttgart in the morning to go to a significant physics event in Lindau, Switzerland, on the way home. Dr. Heid wanted to serve us schnaps as we were leaving, the last thing I wanted before a long drive. This event was sponsored every year by a wealthy count, and the subject area shifted from year to year, that year was physics. All physics Nobel prizewinners were invited and their expenses paid. If they gave a talk, their spouse's way was also paid. The audience was primarily German physics graduate students who then had an opportunity to mix with the Nobel Laureates. There was some other audience, including current Humboldt prizewinners whose expenses were covered. It was a beautiful site, a high mountain lake, the Bodensee or Lake Constance. Ivar and Inger Giaever were also there so we had a nice reunion.

I heard the talks, all by Nobel Laureates, and thought they were almost all horrible. They sounded like fund-raising pep talks about the speaker's accomplishments. Ivar's was different. He talked about the problem he was thinking about, and why it seemed important, and then how he went about setting up experiments to try to learn what was going on. It seemed exactly appropriate for the occasion. I was very proud of him, and of the GE Laboratory which had fostered this kind of approach. I had gin and vermouth with me which we could use to toast his talk before we all left. I gave the remaining gin and vermouth to Jan Taucs, who was returning home, and he looked happier than I had ever seen him.

We returned to Stuttgart for a shorter stay in 1989, also sponsored by the Humboldt Foundation as a follow-up to the earlier visit. We again found an apartment in Botnang but this time in a high-rise building on Furtwanglerstrasse. This time we didn't have the Humboldt BMW but I answered an ad in the local paper and bought an older sedan which served us well. Ole's group had changed quite a bit but the older members were mostly still there. One of the new young ones was Jan Zaanen, a Dutch post doc who came around every noon calling out "Mahlzeit!", time for lunch. Another was Igor Masin, a Russian who had an apartment in the same building as ours in Botnang, with his wife. Igor often rode with me to the Institute and he was quite something. He had a Russian joke for every occasion. One I remember was about a

Russian immigrant living in Brooklyn. He was driving to work with his radio blaring music, going in the *wrong* direction on the Bronx Expressway. The music stopped and the announcer said "Anyone on the Bronx Expressway watch out! There is some idiot going the wrong way." The Russian laughed: "An idiot? There are thousands of idiots going the wrong way!" Igor also knew the area better than we did, and introduced us to the store "Big", which was like Sears and Roebuck and had everything.

I think it was also on this trip that we learned about Besenwirschafts, perhaps from a Bavarian coworker with Manuel Cardona. Private wine-makers could set up tables in their homes and sell the new wine in spring. Everyone considered it a bargain and sat at big tables enjoying themselves and getting acquainted. At one of these we came to know Werner and Susanna Kling from Feuerbach, another suburb of Stuttgart. We ended up seeing them regularly and at some point they decided that we were ready to be *par Du*, speaking to each other in the familiar, *Du* rather than *Sie*. Werner was of the old school and this required a ceremony, meeting together, linking arms, and having toasts — quite a novel experience for us. We learned later that there are certain respon-sibilities of being *Du*. We were supposed to always remember their birthdays and such things that are associated with being in the same family. We've seen them whenever we return to Stuttgart, but probably not lived up to what might be expected from *Du* friends.

During this second time in Germany we took a trip to Gronningen in The Netherlands. We went by rail from Stuttgart to Gronningen with five times umsteigen — changing trains at stations between. They were short stops and we were amazed that every one of them worked, which seems inconceivable to an American. From there we went to Amsterdam, staying at a hotel, the Lion's Head, on the main street, Damrak. We had drinks — and I think maybe pot — after dinner and when we returned to the hotel the clerk said: "Go to your room and turn on the TV. The Berlin Wall just came down!" It was true and a very exciting night because of it, leaving us a fond memory of Amsterdam.

Soon after that we had a visit from Greg Allen and Foo. They stayed with us in our apartment on Furtwanglerstrasse. We took a trip together to Kolmar, an old village in Germany near Strasburg and stayed over there. I remember the train ride there. At the Hauptbahnhof, Greg went off and got a fish sandwich with the head and tail sticking out of the bun at each end. Greg and Foo also wanted to go to Berlin to see where the wall had come down. Greg took the little hammer I'd brought from home and a screwdriver, and used them to chip a number of pieces off the wall, with the paint still on them and we ended up with one or two pieces. There was not much left of the screwdriver.

There were discussions in our group in Max Planck about the effects of interactions between electrons which are usually ignored. I stayed out of it because I was sure I understood them, but in much different terms than they were thinking. Discussion in such circumstances never seems productive. However, Igor Masin came to talk to me on the subject, so I explained my outlook to him. He was quite enthusiastic about my view and told Ole Andersen about it, saying I should give a talk to the group. This somehow made Ole nervous and he said we couldn't do that till Volker Heine was visiting in a few weeks. I then gave my talk, and the response from all, including Volker, was mild. I thought it had no impact, but I believe Jan Zaanen in the group later made quite a name for himself, particularly in the Netherlands, publicizing what to me is exactly this view. He called it $LDA + U$ (Local-Density Approximation with an added electron repulsion, U).

During this visit, we also began enjoying ballet and concerts in downtown Stuttgart. Lucky could pick up the tram on Furtwanglerstrasse and ride down to the Hauptbahnhof, walk past the Turkish workers in front of the station and up Königstrasse to buy tickets. She was becoming imbedded in the Swäbian culture of Stuttgart. Waiting at the tram stop she noticed other women looking her up and down, particularly at her shoes. They may not have been of the standard Schwäbisch type.

We made a third visit to Stuttgart for three months, again supported by the Humboldt Foundation. That was not usual but at various Humboldt affairs we had become quite good friends with the General Secretary and he was happy to arrange it. This time we had an apartment on Kaindlstrasse, the street on which Ole Andersen lived, a short walk from the Institute. Rick and Linda visited us on this trip, staying at an inn in Bünau, just uphill from the Institute. From our apartment Lucky could walk through a field to an S-bahn station with a train to the Hauptbahnhof, so our concert going restarted. The apartment was also a short walk to a local Kneipe, Fritz und Ilse's, in Büsnau, which was quite a find. There were two or three tables but the few customers sat *an der Theke*, at the bar, and we became acquainted with them. It seemed to be the custom that at the bar you spoke with each other in familiar German, and we were comfortable with that. I remember when I studied German in high school and we learned familiar, as well as formal, German, I thought there was no way in the world that I would ever have use for the familiar. But I was wrong, it was very useful here.

It was universal in bars in Swabia, and maybe all of southern Germany, to have one special table, the Stammtisch, where only locals sit. It would have some small tower decoration to mark it and if a stranger sat there it was pointed out that he should sit elsewhere. It is a nasty custom, leaving visitors sitting around the edge, watching the locals enjoy themselves, but we had bypassed it with Fritz and Ilse. They had a Stammtisch but no one sat there. Not only that, if we went into another bar in the area, one of our friends from Fritz und Ilse's would be at the Stammtisch and invite us to join them. That was *really* being imbedded in the Swabian culture.

During that third visit to Stuttgart we also became familiar with the Sportplatz, which had a bar and restaurant, in a valley between the Institute and Büsnau. We sometimes had dinner there and sometimes just a drink. At the end of our third visit, we talked with the owner of the Sportplatz and arranged to have a party with all of our friends at the Institute. The party went well, and even included Alfred Seeger whom I had met on my first trip to Germany, but there was some problem about the charges at the Spotplatz. It is the only time I can remember getting into an angry argument in German, but it ended calmly. We have returned to Stuttgart a number of times for short visits since then, once even staying in the Institute guesthouse, again a short walk to the Sportplatz. Fritz and Ilse had closed down and their space was absorbed by the neighboring Aldi supermarket.

About the time I got my PhD, there was a student, Jim Phillips, getting his degree at the University of Chicago. He went to Bell Labs when I went to GE. He did theory much like my research but centered on semiconductors rather than metals. He was quite a flamboyant character and seemed to feel a competition between us. Some years later, there was a professor in Germany, Joachim Treusch, who was aware of this and thought it would be interesting to have the two of us give adjacent invited talks at the German Physical Society Meeting. He made the invitations including offers to pay the transportation, but Jim was unavailable the first year. We both agreed to talk the next year in March in Münster. He wanted to speak first and I was sitting with Manuel Cardona. To everyone's surprise Jim read his talk in German, starting with this being a German meeting, he thought it appropriate to give his talk in German. I learned later he had written it out in English and had it translated by Werner Weber, a German visiting Bell Labs. Manuel said to me: "My God, this is awful!" and indeed since Jim didn't speak German it didn't sound very good, but Jim was trying one-up-man-ship. I had a half hour to think about how to respond. When my time came I got up and said in fluent-enough German that I was astounded to hear Jim give his talk in German because I didn't know he spoke German. That brought quite a laugh because he obviously didn't know any German. However, I said, you all understand English much better than I speak German, so I think you will understand me better if I give my talk in English. Jim had to smile though he had no idea what I had said. I think my talk went very well. After the session, Treusch took me out to dinner with a number of his students and then bar hopping and much too late at night at Die Schwarze Schaffe. Thus ended a very strenuous trip.

Even before our first trip to Germany we started having Oktoberfests for our friends, good sized parties of perhaps 50 people. We would have a small keg of beer, and lots of wine, and wurst which I cooked in the barbeque, and sauerkraut. We did them annually in October, but as time went on, it slipped a little, Novemberfest, Dezemberfest, and into the next year. By 2014 it had slipped to March, a Märzfest, and that was the last one.

I was periodically invited to conferences in Europe, and whenever anyone offered to pay my expenses I went, usually bringing Lucky along. Often the Giaevers were in Norway and then we would stop on the way to visit them. Once we went with them north of Oslo to a mountain lake, which they called a fjord, and I remember a delicious meal with mushrooms. I learned after that Ivar had collected them in the woods, which would have made me nervous if I had known. More usually they were staying in the family's summer house on Oslofjord, which we would call a river. To me fjords are the mountain inlets on the west coast. His family had made much use of it during World War II when Norway was occupied by the Germans. The house is near the town of Tonsberg where the fjord is widening to the south, and where I bought my Norwegian sweater. It is a comfortable house right on the water and with a guest house where we slept. Ivar let me know that booze was very expensive in Norway so I always arrived with bottles, which we made use of. The water was cold, but

Lucky and Walt with Inger Giaever at the house on Oslofjord.

of course Ivar and I went swimming every day. We used masks because, surprisingly to me, there were octopus floating around, apparently brought up by the Gulf Stream. Once we went fishing, Ivar hanging a line over one side and me over the other. Ivar caught a few fish but I never seemed to get a bite. We also went hunting, again with masks, for mollusks on little islands in the fjord. The Giaevers returned often to the house, but in recent years decided to sell it. We went over in one of the final years, but it has turned out that their daughter Guri and her husband were able to buy it, so it continues in the family.

In addition to our many visits in Europe, we had many visits by Europeans at Stanford. Whenever someone wrote to me asking to visit, and had funding to sponsor the visit, I said yes. The one exception I remember was a young physicist who had funding for a year. She wrote from Upsala, Sweden, and I knew John Wills had been there recently so I called and asked him if he knew her. Wow! He did and if he saw her coming he'd hide. She didn't understand anything, but was overly tenacious and eager to talk. So her, I didn't accept. I heard that she went instead to Cornell with Neil Ashcroft. A year or so later I ran into Neil's wife and asked how it had been. She groaned and said it had been a hard year. I guess I was just lucky!

The people who did come worked out well. An early one was Alexander (Sasha) Rusakov from Moscow. It was an Eastern European Exchange Program and an Estonian came at the same time so they shared an apartment while here. These were stern times in the Soviet Union and Sasha had to be careful. He got a copy of *Dr. Zhivago* by Pasternak, forbidden in Russia. He kept it hidden and only read late at night when the Estonian was asleep. We all got on fine, but it had little to do with physics. When he returned to Russia he began promoting a crazy idea for high-temperature superconductivity, which never went anywhere. I later heard from my friend Genrich Krasko that Sasha was a KGB agent, but I found it hard to believe. We did cross paths at some later time and he pretended not to recognize me.

Another visitor from Eastern Europe was Friedhelm Bechstedt, from Jena in East Germany, and we have stayed in touch with him and his family. Lucky and I actually visited them in Jena just a week before the wall came down in Berlin, and we have returned to see them since. He knew a Kneipe in Jena that he thought we would like and we did. It was a jovial place where we got to know all the others who happened to be at our table.

On one of Bill's German stays, he travelled through Jena and stayed over a night with the Bechstedts. The son Andy was quite taken with Bill and gave him his favorite book, *Heinrich Crossmanns Grosse Fahrt*, about a Hessian soldier fighting in the American Revolution, and then deserting to America. We both read it. There was a sad ending: when the wall came down Andy and a friend set off on a bike ride through Europe, and Andy was killed in a traffic accident.

On one occasion a group of Russians were visiting Stanford. I talked with a couple of them and we went to lunch before they were to go to SLAC for an afternoon of conversations there. At lunch I told the two that if they wanted to skip the afternoon session I could drive them to the beach. They were delighted and agreed. Then a couple of others heard and wanted to join us. Pretty soon the entire delegation wanted to go to the beach. The organizer arranged a very short chat with a few people who came in from SLAC and with the cars of a few other Stanford faculty we all drove to the beach. The Russians got out, ran down to the beach, took off their shoes and waded into the water. Then each threw a coin or two out. It must be a Russian tradition. I thought we'd made it a successful visit.

Another visitor was Winfried Mönch from Duisburg, West Germany, who came on an Army Research Projects Association (ARPA) exchange grant between Stanford and Germany which had been initiated by Bob Huggins. These German visitors came every year and I took it upon myself to be unofficial host to them, particularly after Bob was away on an extended stay in Germany. I was particularly impressed with Winfried when I asked him a German question. We had been with a visitor in the neighborhood when he flicked a spider from his jacket, and he mumbled a poem in German about spiders in the morning, but I didn't catch much. I asked Winfried if he knew of such a poem. Not only did he immediately recite it, but he went on to tell the history of it. It was originally about spinning yarn, that if you do it in the morning, it's a job and stressful. If you spin in the evening, it's a hobby and is just relaxing. It has come to be interpreted as meaning that seeing a spider (*die Spinne* is German for spider, and *spinnen* means to spin) in the morning is bad luck, but seeing one in the evening is good luck. I don't know what to make of the last line, spinning at midday brings luck on the third day, but I think every German youth knows the poem. Also it has been translated into French, with *araignée* for spider, giving no clue to the history.

I kept track of this ARPA program at Stanford, and with Bob Huggins gone it was taken over by others in Materials Science until finally under Stig Hagstrum the money was being used to support his postdocs, with no connection to the original purpose. I managed to wrench control of it and return it to bringing Germans. At the time there was enough money for two more years. For the first year I picked Klaus Ploog, a professor in Berlin. The arrangement with Volkswagen was that he would choose a younger colleague from an industry to come, also with his spouse, at the same time, which he did, Henning Riechert, and we found apartments for both couples in Palo Alto. The second year I brought Hans Queisser from Max Planck in Stuttgart. I had known Hans for years. He was one of the German physicists brought by Shockley to Palo Alto, as described in Chapter IX, but had since returned to Stuttgart. He brought Ines Stolberg from Jena, working at Jena Glass in East Germany. She was an interesting choice since East Germany had little connections with the West till about that time. Her husband Klaus accompanied her for part of the time. She worked in another group at Stanford, with Fabian Pease, but we saw them often and she returns every year for a meeting in San Jose when we generally see her. Their visit ended the Volkswagen program at Stanford.

Walt canoeing on Fallen Leaf Lake.

Chapter VIII

EMPTY NEST

When Bill and Bob left for college we offered the Annex for rent to students at the Stanford housing office and it was occupied continually for some thirteen years by a series of students. Lucky got the rent money and it worked well. She had worked for a number of years at Stanford Admissions but this was interrupted by our first sabbatical in Germany.

Our first renter put up large, white Chinese lanterns, which are still there in the Annex. Mark Segal, a friend of Bill's at Stanford, was an early renter whom we still see occasionally. Another friend of Bill's was Mike McFaul, who lived there while Mike was doing graduate work. He was subsequently U. S. Ambassador to Russia and now a professor at Stanford. Another was Tom Dobrocky, an ex-roommate of Bill's, who held a nice graduation party with his family in our backyard. An early pair of renters were Susan and Sally who visited us in Stuttgart. We didn't think the Annex was suitable for two people, but they persisted and seemed happy there. Another, Eric, was in the Annex when we were away on sabbatical and rented the house to a visiting professor from Seattle, who sublet to a colleague of his from Heidelberg and his wife. We actually traded cars with the Heidelberg man, picking his up in the airport garage in Frankfurt and delivering it to Heidelberg at the end. In spite of all that, we never met him. He surprised the Seattle professor by showing up here with a newborn baby. The three lived in the Zanana (third bedroom). Erin Palm was one of the last renters in the Annex. We were invited to her wedding on a Caribbean island, but didn't go. We saw them when they moved back to the area.

After we were finished with scouting, the boys continued to backpack in the Sierras, and occasionally I joined them. Each time it seemed a little harder, and we have a picture from my last trip into Desolation Valley, in the Norwegian sweater.

Walt's last backpack in the Sierras, in the early '90s.

Between our sabbaticals in Germany, we took three-month sabbaticals in New York City, as a visitor to the IBM Research Laboratory in Yorktown Heights, New York. Generally, I seemed to end up taking three-month sabbaticals every three years, rather than the usual one-year sabbaticals every seven years. It worked for us. The first time at IBM we drove east, stopping on the way to visit Rick in Los Angeles, friends in Los Alamos, and Texas. We stopped to see my Uncle Richard and Jerry in Louisiana and went through the deep south and up to New York.

Finding an apartment was not easy. We had hoped to stay in Greenwich Village, but didn't find a suitable rental there. We ended up subletting an apartment on East 35th Street. After that first rental, midtown-east was our neighborhood. The apartment was a mess, but it had a view of the Empire State Building. The bed was a mattress on the floor of the bedroom, with a ridge of boric acid around it to keep cockroaches away. We cleaned it all up and filled one closet with the rubble. We put out multiple roach traps, and it proved comfortable. Also there was a bar at the corner of 35th Street and 3rd Avenue, The Fireside Inn, which became our local. It turned out most convenient to keep the car in Croton-on-Hudson in a parking lot at the train station. It was a nice walk from our apartment to Grand Central Station where a train took me to Croton. I could then drive to the Laboratory a few miles away. Many days I even had a rider whom I had met on the train. He was working on voice recognition programs for IBM which seemed alien to me then, but has become commonplace now.

I felt very much at home at IBM. Sok Pantelides, my ex-postdoc, was head of the group and many of the others in the group were old friends. One in particular was Jerry Tersoff, whom I had known when he was a student of Marvin Cohen in Berkeley. He had followed up an idea of Volker Heine's about metal-semiconductor interfaces, and was widely recognized for the associated "metal-induced gap states". I thought it was nonsense and we talked about it at length. One afternoon I came up with a nice test which I thought might be conclusive. After I made the argument I said: "Wait! It's coming out on your side!" I said it's time for me to catch the train, which it was, and that we'll talk tomorrow. We did, and co-authored a paper on it. I still think it's a terribly misleading concept, but it's nice that we could find a common ground.

Living in Manhattan was wonderful. We had dinner out every night, and managed to do it reasonably cheaply. We picked up a book at a street fair which had discount coupons for many restaurants and we explored those which sounded interesting. One was for an Afghan restaurant in St. Mark's Place in the East Village. When we got there it was a dumpy take-out restaurant, but next door to it was an interesting-looking restaurant, Khyber Pass, also Afghan, which we went to *without* a coupon. It has been a regular for us since — across from Bull McCabes bar with which we are very familiar. Most weekends there were street fairs where we'd also see another part of the City, and street fairs are always good for lunch. The theater was also more than just theater; Lucky would go to the line in Times Square for cheap tickets, TKTS, and usually make a friend on the line. We'd have dinner near the theater, and after theater have drinks in the neighborhood before taking the subway back. At the end of the stay, John flew out and with Todd Handy drove our Mustang back to California. We flew back and the trip turned out to be our last drive across the country.

For our next stay in New York, we found an apartment on East 54th Street at Second Avenue, operated by the Hart-Parker agency, suggested by one of Lucky's friends in Admissions at Stanford. It was right next to the old El Morocco Night Club, which had been one of the best-known nightclubs in New York, but was on the way out. The apartment building was managed by Teresa Swenk with whom Lucky became a close friend.

We needed a new local bar and Parnell's, at 1st Avenue and 53rd Street, became our long-standing favorite. If we ever went back to the Fireside Inn, it was all going to be new and unfamiliar people. At Parnell's,

Lucky and Walt having dinner at Parnell's, 2017.

we always have friends and a bartender we know, the last several years — Ambrose Blain. The 3rd Avenue subway station is close by, and a good way to get to the theater district. It became a tradition with our boys to come and visit while we were there and we were delighted to have them learn more about the City. At Hart-Parker we always got another apartment for them, and Teresa charged us very little. We would typically have dinner together, and they had things to do in the evening, maybe joining us later at Parnell's. Sometimes we would take my favorite excursion, the 2nd Avenue bus to the Staten-Island Ferry, to Staten Island and back, and the subway home. Eventually Hart-Parker came to an end. Teresa was called in by the owner and told that she needed to pack her things from office and she would have to leave. I had heard that this happens, but was shocked. We found a very similar arrangement at Global Quarters, around the block — you can see the Hart-Parker building from the kitchen window. It wasn't hard to adjust. That has been our home-away-from-home ever since. We have gotten an extra room for visiting boys, but now they prefer to find their own place. They know their way around and it works well.

During each of these mini-sabbaticals we rented our Stanford house out, usually to a visiting professor. On one occasion, the professor who had arranged to stay found a house in Menlo Park which could take him for the entire year he was to be there. He was very apologetic, but wished to cancel ours at the last minute, to which we agreed. We inquired at the Stanford Housing Office and found a set of five undergraduate girls who wished to rent. We had some misgivings but agreed and Lucky met with one, indicating that it was a quiet neighborhood and noisy parties would be frowned upon. The girl said: "Oh, don't worry! We're all in electrical engineering and we are nerds." We don't know how they arranged accommodations for all those people, and we heard that at some point a male student moved in with one of them. It seemed to work out well, and each month I received an envelope with five checks, of varying amounts but adding up to the agreed rent. They were very well organized and I guess meal costs were mixed with the rent.

Front row, Mark. Jennifer, Lucky, Walt and Rick. Second row, John, Ace, Bill, cousin Annie Harrison, Bob, and Linda.

By this time the boys were clearly men. Rick had married Jackie Sutphin, who had a three-year-old daughter, Toni, and Jackie then gave birth to Ace. After two years, there were problems. Jackie spent her time demonstrating for Scientology, which Rick was in favor of, but it left him trying to take care of the family and earn a living at the same time. At some point, he headed off for Colorado, hoping the family would join him. They might have, except with unreasonable conditions such as monthly returns to LA. Rick met Linda and soon they were married. They applied for, and adopted, two children from Russia, Mark Yuri Harrison, about 11 years old, and Jennifer Veronica Harrison, five years younger. Rick and Linda needed to go to Magadon in far eastern Russia to pick them up. Linda and Rick managed to bring them up through high school, which was no small feat since so many Russian adoptees have serious problems adjusting to American life. After high school they each, in turn, moved out and are on their own. Jennifer married and had a baby. She had serious problems with drugs and left the baby with Cole, her husband. For a while Jennifer had a job as a nanny and was stable but the last we heard she was asking for bail. Mark earned an associate degree in Colorado and now works there in construction, living in his own apartment.

Rick's son Ace remained in LA with his mother until she remarried and moved to a farm in northwestern New York with her husband, Gary Fleckenstein. In the mean time, Toni got married and had a daughter, Violet. When that marriage came apart, Toni and Violet moved in with the Fleckensteins in New York. Ace remained a while in Los Angeles but then also joined the Fleckensteins. Gary had quickly accumulated quite a family, but it seems to be a happy one. The farming aspect is quite limited, a number of cows that need to be moved from pasture to pasture, and Gary and Ace are handling it. Ace comes west about once a year to join us for a stretch, sometimes including a beach stay. He has contact with Rick only on such occasions, if Rick happens to be here.

John was working as a silk-screen printer in San Francisco, first printing T-shirts and skate boards, and then fashion blouses. In that last job the company was hiring another printer so they put an ad in the paper, "Silk-screen printer wanted". There were many responses and John thought the girl they hired was very good, but it paid minimum wage. It's a tough business. He tried various other jobs, one with Veritable Vegetable, first in the warehouse and later driving trucks. Driving trucks continued later with Alvarado Street Bakery, but evolved into a managerial desk job. That change was very fortunate since soon after that Alvarado cut out the truck delivery service. Earlier John had a companion, Eric Newton, and they shared an apartment in a new development near the Bay Bridge. That relationship came apart

John and KK.

and he then tied up with Ken Downey, now KK, and that has lasted. At one stage they bought a record store on Polk Street in San Francisco where there were a number of other record stores. One-by-one they closed down, and theirs went too. It was the end of an era. KK now acts as host in the classy Market 1 restaurant in San Francisco.

Bill was *immer unter Wegs*, he had spent time in Stanford in Germany, in Berlin, and he returned with a Fulbright to the Technische Universität in Berlin. He came back to San Francisco and joined a construction engineering firm. One of the other employees, Sarah Leong, left to form her own firm and soon Bill joined her. It was a little like being an independent consultant and has lasted ever since. He bought and fixed up a house in the Excelsior, south of the City. I was skeptical when I first saw it, wallpaper peeling off and I believe a hole in the floor, but he made it into a comfortable and attractive home. He continues to travel to Berlin regularly. He was asked by a couple in Oakland, Mya and Whitney, to be a sperm donor and agreed — twice. From that he has two loving sons, Luka and Kasper Thorniley, whom he sees weekly. More recently he married Geng Thimsuren, a young man from Thailand, and they are getting on fine.

Bill and Geng.

Bob had been working for the Bank of America in San Francisco and transferred to the office in Palm Springs, where he remained. He bought a house in Cathedral City, adjacent to Palm Springs, and is still there. At some point he retired from Bank of America, and went back to work, mostly as a caregiver, but for a while, with a funeral home. Then he took a job with the U. S. Mail as a carrier and has continued in that job. Happily they are flexible on allowing time off and we see him here regularly.

All of our sons have taken music lessons on some instrument of their choosing, as had I when I was young. In my case, my father had suggested the accordion, and that was my choice. It didn't last and I stopped lessons by the time I went to high school. Sometime in late high school, our gang was in a bar where an accordionist was playing and I wanted to give it a try, but I put it on upside down and it felt totally unfamiliar. Then either in late high school or early college I took up the ukulele. I went to some pawn shops in Toledo to find one and came upon a banjo-uke, which I bought. It was louder and worked well in the bar at fraternity parties. However, when a friend offered to trade his fine Martin Ukulele for my banjo-uke, I couldn't resist, and I have the Martin to this day. After college the occasion to play it didn't come up so often, but occasionally at beach fires I would bring it out. Then one year at Stanford, Lucky gave me a banjo and lessons for a birthday present. I worked at it till I thought it sounded pretty good, and then I recorded my repertoire on a Dictaphone. When I played it back, it sounded so terrible that I just put the banjo away and forgot about it. Maybe a Dictaphone is not a fair test. In any case, after a few years I thought of it and wondered if I still knew how to play. I took it out of the closet and sat down and out came the music. It was a great contrast to the experience with the accordion. My mind didn't know the chords but my fingers did! I liked the feeling and I kept it that way, and I *never* recorded it again.

I remember one weekend sitting on the beach at Rio del Mar by myself playing the banjo. The next week I talked to Ole Andersen, who was visiting Xerox, and he asked me if I'd been on the beach. He and Sanne had gone to that beach and had thought that the man in the cowboy hat with the banjo looked like me, but it couldn't have been! I also occasionally picked up the old Martin Ukulele. It got old and cracked in back and I have a newer one, not a Martin. I guess Bill took up the guitar again more recently and took up the mandolin, but it didn't last and I don't think that the other sons have gone back to their instruments.

On some occasion, I was discussing multidimensional space with a visitor and I mentioned it at home. Our son John got interested in the subject. After a little talk about a four-dimensional cube he said: "I think I could make one!" I didn't know what he meant, but he went into the garage and from some sticks and glue he constructed a model. I believe

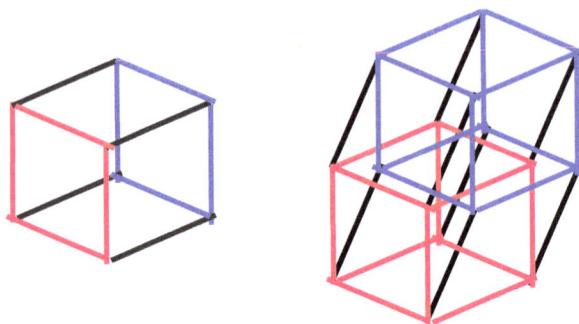

On the left is a three-dimensional cube, projected in two dimensions. On the right is a four-dimensional cube projected in two dimensions. Projecting either in one dimension leads to a simple straight line.

it was indeed a three-dimensional projection of a four-dimensional cube. That seems to me something much beyond artistic talent, which I knew he had. Above on the right is a graphic representation of a four-dimensional cube. The model that John built is still hanging in my office.

Lucky came in the middle of each visit at Sandia Laboratories, and later Los Alamos, usually coming the first Thursday and flying out the next Monday. We stayed in Santa Fe while she was there, then Sunday night in Albuquerque and I returned Monday morning to Los Alamos. We stayed at many motels in Santa Fe but in the end settled on Garrett's Desert Inn, a short walk from the plaza. It was also just a few doors down the Old Santa Fe Trail to the Pink Adobe restaurant, where I had had dinner on my interview trip at Los Alamos in 1956. It also had an appealing bar attached which felt like it was outdoors.

One year when I returned to Los Alamos on Monday, I was missing my office key. The secretary said I'd better find it or they would have to redo all the locks in the building. Shocked, I returned to all the places we'd been and finally I spotted the key, imbedded in the dirt next to the bar in the Pink Adobe, and imbedded in my mind forever after. We often had dinner at the Pink Adobe, or the Coyote Café, or at Pasqual's for New Mexican specialties. Another favorite spot was the top of the La Fonda Hotel, where we could have drinks looking out over the town. There was also music in the La Fonda bar, and sometimes dancing. Bill Hearn played guitar and his wife, Bonnie, sang country-western songs.

On one of these visits, John joined us and we went rafting on the Rio Grand River near Taos. We had a two-man raft and I assumed I would be in the back to steer. John suggested he go in the back, and I realized that

John and Walt rafting on the Rio Grand.

Walt and Chuck in 1985.

he was right. There comes a time when the younger person should take over. For rafting this was the time. The picture shows the busiest part of the trip.

My stay in Los Alamos ordinarily coincided with the season for Hatch peppers, from Hatch County near the Mexican border. We would buy a 35-pound sack and I would take them back in a duffle bag on the plane. I roasted them on the barbeque and we pealed and froze them for later use. They are an important addition to a cheeseburger. I remember one year when a Los Alamos intern in T1 heard us talking about Hatch peppers and then bought a sack. He asked Galen what you do with them. And Galen replied: "You don't *do* something with Hatch peppers, they are a way of life!" I agree. The Santa Fe Flea Market, north of Santa Fe, is quite a special one. We bought our H-bar branding-iron candle holder there (reminding me of $\hbar$, the h-bar symbol of quantum mechanics), lots of turquoise, and there are always the tamales. Also in Santa Fe is the Nambé store, where Lucky added to her collection of Nambé ware, a quite impressive one. We visited Don and Corinne Carlson when they were living in Taos and Ann Carlson when she lived by the river south of Taos, near where we had been rafting.

I also had a long-standing relation with Salt Lake City, Utah. Craig Taylor invited me to spend a week in the Physics Department in the University of Utah in the early eighties. I did and made good friends with a number of the faculty there. I stayed for a weekend of skiing at Alta, which was wonderful. I was invited again every year, and each year I invited one son to join me for the weekend of skiing, starting with Rick. It was a unique opportunity to spend an extended time with a son each year. One year, in the middle of this period, Lucky joined me and a couple of sons in Salt Lake City. Eventually Craig Taylor moved to the Colorado School of Mines in Golden, Colorado, and I continued weeklong visits there, but we didn't continue the ski tradition. We had another tradition in Craig's group, which was carried forward from Salt Lake City: I would go with a large group of graduate students and post docs to the Coors brewery for beers and dinner, covered financially by Craig.

In 1994, we celebrated our 40th wedding anniversary and decided to have a sizeable family party. We invited Chuck and Carolyn out from Toledo for the occasion and arranged for them to stay in the Creekside Motel in Palo Alto. Rick and Linda came from Colorado with Mark and Jennifer. We also brought Bill and Martha Rose, and Sara, but a little later for a smaller celebration. With the first group we had a sizeable dinner with all of the family and a number of friends. On a separate day we rented a forty-foot limo for a trip for the family to San Francisco. It was a rare occasion to have so much family together.

During the fifteen years or so after Bill and Bob left for college and before I officially retired, my research continued in several directions, but almost all centered on electronic structure. Some was jointly with Ed Kraut and others at Rockwell, with Galen Straub and others at Los Alamos, with Craig Taylor and others at Utah, and with students and post docs at Stanford. It covered simple metals, transition metals, semi-metals, covalent solids (semiconductors), and ionic solids (insulators), pretty much the whole range of solids. The goal in each case was a simple enough representation of the electronic structure that one could use it to estimate almost any property of that material with simple algebraic calculations. It in fact turned out to be possible. This required some input numbers, such as the valence s-state energy and valence p-state energy for each element, but these could be tabulated in a separate table. All of these years were required to achieve this goal which I had to some extent envisaged at the outset.

Of course I put this together as a book, *Elementary Electronic Structure*. It was like a new edition of my *Electronic Structure and the Properties of Solids*. I had become acquainted with World Scientific Publishing Company in Singapore in connection with a number of conference proceedings. They were willing to publish books more cheaply than companies such as McGraw-Hill, and Freeman, and I was much more interested in wide distribution than in making money. They were paying lower percentage on a lower-cost book. They published mine in 1999 and a revised edition five years later. I told of designing the cover for this book, and the flaw in the pattern, in Chapter V.

During this period after the boys left I mostly taught electronic-structure courses, but sometimes group theory. I thought about volunteering to teach quantum mechanics to the Physics Department graduate class. However, they had a fixed syllabus which wasn't at all what I thought was appropriate. There were long segments on the formalism for quantization of angular momentum, the relativistic theory of the electron, and fundamental particles, and I had no interest in teaching any of those. My feeling was that most faculty thought that these were basic to quantum theory and it was proper that every physics PhD should go through them. I thought that quantum mechanics was much too important to be ruled by tradition, and that we should teach what is really useful. I felt supported in that view later when a physics graduate student commented that he didn't understand the physics when he took the Physics Department course in quantum mechanics, but assumed that maybe later it would become clear to him. There had also been a course on quantum mechanics for electrical-engineering graduate students organized by Bob White, but currently not being taught. That looked perfect so I volunteered to the EE chairman. He said great, but there was a new man, David Miller, joining the EE department and they were thinking of him for the course. I said no problem, he wouldn't be teaching it the first year as he got settled and after that I would give up the course whenever he was ready. All that worked, David happened to move into the house next door to us, and didn't take the course for two or three years.

I was quite excited about this new course. I had the idea of saying at the start that there was only one assumption in quantum mechanics, *Everything is both a particle and a wave.* I wasn't sure that it was true that there were no other assumptions, but I think it is true enough and with good emphasis. I followed that up with a series of applications and simple calculations of properties of atoms and solids, based on this premise. It was a major departure from the standard approach of looking at the solutions of Schroedinger's equation for simple, idealized systems. I couldn't tell if the students, mostly from Materials Science and Electrical Engineering, shared my enthusiasm.

Later I substituted for a few lectures in the Physics quantum course. I told them that I would give the most useful parts of quantum theory in these few lectures. And they seemed not to be impressed. They were more interested in Dirac theory, the relativistic formulation of quantum theory which they were about to begin. Maybe I was shouting into the wind, but I gave a few lectures on how one can actually use quantum theory for answering interesting and useful questions. While teaching my own applied quantum-mechanics course, I organized my lecture notes into a book, *Applied Quantum Mechanics*, which I published with World Scientific.

I remember at a meeting later running into a young physicist whom I had met in Stuttgart, and he told me he was teaching a course using my book as a text. He thought it was great, but he was puzzled that the symbol $\hbar$, the central quantity of quantum mechanics, was so awkward looking. He remarked that the publisher must not be very competent. I assured him that it was my fault; I sent the publisher camera-ready copy that I had produced. My text file didn't have an $\hbar$, and I had tried to construct one from the software that I had. I guess I didn't do so well at that.

After a couple of years David Miller was ready to teach the course and I went on to something else. He didn't like my text too well and within a few years wrote his own text. He said that we were the only case in history of two next-door neighbors writing texts on applied quantum mechanics. I expect that was a safe statement.

When I was chairman of the department, I decided that it was inappropriate for people to stay beyond 70 years old because it prevented the department from hiring someone new. When I was coming up to 70, the dean announced that there would be no new hires for a while, so that pressure was gone and I stayed on teaching until I was 71.

We had earlier brought family from the east to celebrate our 40th wedding anniversary in 1994. To celebrate my retirement in 2001 we hosted a very large *Emeritusfest*. We bought monogramed wine glasses as favors and rented lights and tables and chairs for about 100 people in the backyard. We brought Rick and his family from Colorado and had a large representation of Stanford people. We cooked wurst and kraut as for our Oktoberfests and also had music on the side by Blaise Martin's ensemble, that our son Bill had sometimes played mandolin with. They played old-timey music, resembling blue grass but more informal, and it was very pleasant and unobtrusive. We hired some students to help out and it all worked quite well. The party lasted well into the night. I remember Frank Cooper late at night showing his prowess at yanking a tablecloth off the table without disturbing the glasses on it, almost.

We did one more of these large fests, following the same format, in 2004 to mark our 50th wedding anniversary. For this one we hired an old-time music group that Blaise knew of, to again provide unobtrusive music in the background. At some point in the evening, Rick's Jennifer tried on Lucky's old wedding gown and made a memorable appearance in it among the dining tables.

I had not thought to make any kind of public announcement during the dinner, but our son John thought it appropriate and got up to toast the occasion and introduce our visiting family and friends at the central table. It was a nice gesture which I should have thought of. A little later we again had MR and Bill Carley over, with Sara and her husband Todd Morris and son Blaze. That visit included a trip to San Francisco and time at the swimming pool. John shone as a swimming and diving instructor.

Chapter IX

RETIRED

In some sense, retirement had quite a small impact on my life. I stopped teaching classes, which freed up the schedule, but my research continued as before. Also of course paychecks stopped but I was receiving required (by the IRS) minimum distributions from my retirement account and it was almost all we needed.

In another sense, it was a whole new ball game. I had essentially accomplished my long-range goal of providing a new and simple understanding of the electronic structure of the entire range of materials, and had documented that understanding in a comprehensive text. I gave talks on this general understanding at various places, including Stanford, but I was beginning to realize that my audiences were not nearly as enthusiastic about this accomplishment as I was. The rest of the world did not have the same goal as mine. Those working in areas close to this were more interested in computer codes which could predict the same properties which I could predict, and probably more accurately. Understanding in the sense I had developed was not a central question to them. It reminded me a little of my early days at GE, when our group seemed to have different motivation from those at Bell Labs, who were considered the leading lights of the times. This wasn't as disturbing to me as one might expect; I had my view of how science should proceed and I would proceed the same way if I had it to do over again.

I did continue clearing up some aspects of the electronic structure of materials, but broadened into other questions, a little like earlier times when we were on sabbatical and I toured Europe talking about the meaning of the question, how long does it take? That's a subtle question in quantum mechanics, not widely understood. Ted Geballe once stopped in the office and said he expected me to be interested in all the new activity on valence-skipping compounds. I did in fact know about them but considered the discussion to be nonsense. I took the occasion to clearly state my simple understanding of the question and published it as a short paper.

I also worked with a Materials Science student, John Jameson, who came with very interesting questions that we could sort out together. John was doing experiments on amorphous materials similar to the glasses I had worked with as a student at Owens-Illinois Glass Company. He found that when he applied an electric field to these materials they would relax to reduce the field as most materials do, but then would continue to slowly relax further. That could be understood roughly in terms of what is called a two-level model, a system which can be in either one of two different configurations, and which can jump from one to the other. In this case, it would be a very large number of neighboring configurations. We were able to understand a whole range of properties accurately with this model. John and I published three papers together using this approach and I included it as the last fifteen pages of the second edition of my book *Elementary Electronic Structure*. One of the leading experimentalists who was also doing experiments in such systems was a physics professor at Stanford, and our approach neatly explained a number of his findings. I was interested and surprised that he was not particularly interested in our explanation of his work.

Another problem which came up was the traditionally intricate calculation of the electrostatic energy of ionic crystals. I found a very simple shortcut which made the calculations quite easy and happily wrote a two-page paper pointing out the trick. When I submitted it for publication, the referee complained that I had not referenced and discussed the many previous analyses on the subject. My whole goal was pointing out a shortcut, not padding it into a treatise. In the end, I won and was allowed to publish my two-page trick. It seemed at the time that nobody had noticed. This could be another case of being out of register with the dominant culture of the time.

Some six months or so after I retired, I had two visitors from the National Energy Technology Laboratory in Morgantown, West Virginia. They were supposed to be supporting the coal industry in West Virginia and were developing fuel cells that could be based on coal. They wanted to hire a post doc who could do computer calculations connected with such fuel cells, and were asking where to look. I could suggest some people who might be promising graduate students, but then said that if they really wanted help with their program they might do better to hire someone like me as a consultant. They jumped on the idea and arranged for me to be a consultant through Leonardo Technologies, Incorporated.

I made a visit to learn what they were about and learned that it had to do with making devices that oxidized methane in a way to convert the energy directly into electricity. That required diffusing oxygen through some manganite compounds that I was familiar with and I undertook a study of these materials and published papers on them. It gave me an occasion to make some trips to Morgantown, where I had earlier visited my student John Wills. It was nice to have a problem to work on and people interested in the results, and also nice to have consulting fees rolling in to supplement my required retirement withdrawals, even if it wasn't necessary. Everyone was happy and it went along for several years. I even have a retirement fund connected with the work which both I and Leonardo Technologies contributed to. Suddenly the Department of Energy which had been funding the activity decided to terminate it. Our friend Steve Chu was actually heading the Department of Energy at the time and it was his decision, but it had nothing to do with me.

I was sorry to see my connection with the National Energy Technology Laboratory end but had extra time and decided to write another book, which was intended in simple terms to describe my general understanding of solids in terms of their electronic structure. It was to cover all elements, shown in the Table. It is in order of increasing atomic number but with different breaks than usual so that it would better reflect the electronic structure of compounds of each element.

Transition Metals   I II III IV V VI VII VIII   *f-shell Metals*

H He Li

Be B C   N O   F Ne Na

Mg Al Si P   S Cl Ar K Ca

Sc Ti V Cr Mn Fe Co Ni Cu Zn Ga Ge As Se Br Kr Rb Sr

Y Zr Nb Mo Tc Ru Rh Pd Ag Cd In Sn Sb Te I Xe Cs Ba La Ce Pr Nd Pm Sm Eu···Yb

Lu Hf Ta W Re Os Ir Pt Au Hg Tl Pb Bi Po At Rn Fr Ra Ac Th Pa U Np Pu Am Cm···

Periodic Table of the Elements, rearranged. In Column IV are the semiconducting elements and in Column VIII are the inert gas atoms, with all the nonmetals between.

Column IV under carbon lists the semiconducting elements. Column VIII under helium lists the inert-gas atoms and all of the nonmetals lie between these two columns. When a nonmetal forms a compound with a metal to the left of Column IV, it is a semiconductor. When it forms a compound with a metal to the right of column VIII, it is an ionic insulator.

I called the book *Theoretical Alchemy*, a term which described a central part of the understanding. Recall that alchemy was originally the attempt to convert lead into gold, change one element into another. It would be starting with lead, in Column IV, Pb with 82 protons in the nucleus, removing three protons to leave 79, the nucleus for gold Au in Column I. The atom would shed three electrons to become a neutral gold atom, though at the time no one knew about nuclei.

In my own alchemy we start with a silicon crystal. Silicon is in Column IV indicating four valence electrons per atom and forms a crystal with each atom having four nearest-neighbor atoms. As illustrated in the figure, each silicon atom has a two-electron bond with each of its four nearest neighbors, the simplest description of the electronic structure of silicon. I then imagine removing a proton from alternate silicon atoms, moving them one column to the left in the table and making them aluminum atoms. We then insert the protons in the nuclei

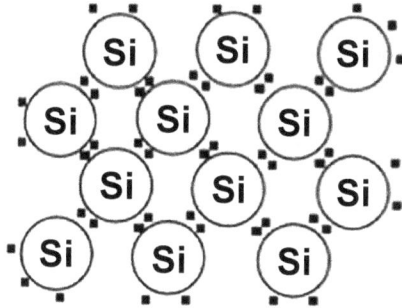

Silicon two-electron bonds. Each silicon atom puts one of its four valence electrons, each represented by a dot, into a two-electron bond with each of its four neighbors. Those left over at the edges are called dangling bonds.

between, moving them one to the right and making them phosphorus atoms, and thus forming a crystal of aluminum phosphide. The crystal structure is the same and the bonds shift a little toward the phosphorus, but the electronic structure is essentially the same set of two-electron bonds. We can then transfer another proton to make magnesium sulphide. We can do the same exercise starting with germanium, directly below silicon in the periodic table, or carbon directly above silicon. It is a simple way to understand the electronic structure of all of the common semiconductors in terms of the pairs of electrons forming the bonds.

There is another set of transformation which describes the electronic structure of insulators consisting of two types of atoms. Note that after I had constructed magnesium sulphide in the last paragraph, I didn't transfer another to make sodium chloride. Sodium chloride, rock salt, has the atoms arranged in a different pattern, a different crystal structure, and has a different kind of electronic structure. To understand this other kind of electronic structure we start with a crystal of argon atoms, inert-gas atoms in Column VIII of the table. The eight electrons fill the "shell" of atomic states for each atom. The last electron is strongly bound to the atom but adding another would put it in the next shell and it would be very weakly bound. The Column VIII atoms are very stable and don't form bonds, which is why they are called inert. We imagine arranging the inert gas argon atoms in a simple cubic lattice, but then remove a proton from the nucleus of alternate atoms, making them

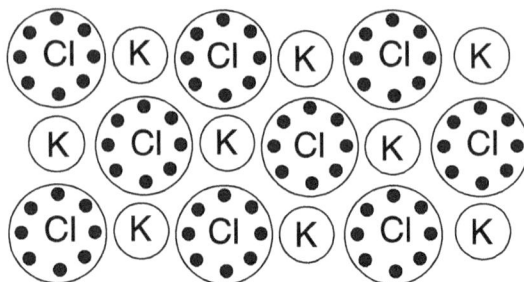

Potassium chloride, KCl. Each chlorine atom has an extra electron making it a chlorine ion and each potassium atom has lost one electron to become a potassium ion.

chlorine nuclei, and add the protons to the others making them potassium nuclei, potassium chloride, very similar to rock salt. It is composed of quite inert potassium *ions* (charged atoms, one electron missing compared to a neutral potassium atom) and chlorine ions (charged atoms with an extra electron). The ions are inert but are held together in the crystal by electrostatic forces. Moving another proton makes calcium sulphide. To make sodium chloride we would have to start with alternate argon and neon atoms. Surprisingly, this sharp qualitative distinction between covalent solids (based on two-electron bonds) and ionic solids (based on full electron shells) is not generally made. Linus Pauling for example put them all on the same "ionicity" scale.

About the time that I finished the book three people from Acorn Technology Incorporated stopped at my office. They were trying to reduce the resistance of contacts in tiny transistors. They went to Jim Harris in EE at Stanford and he said he had no idea what they should do about it, but suggested they talk to me. They did and told me what they thought was their problem. I said that I didn't think that was the problem. They asked what they should do to lower the resistance. I said a thin layer, one atom thick, of arsenic at the contact should help, and explained why. They were delighted and hired me as a consultant, up to 40 hours in a month.

I started meeting for a day each week with the two Acorn employees who were local, Paul Clifton and Andreas Goebel, an Englishman and a German. They were physics-oriented and our discussions were interesting and seemed very fruitful. Between meetings I would work out the consequences of our discussions and got very much interested in the subject of metal-semiconductor interfaces. I was pretty much working full time on these problems, but of course just charging for the 40 hours a month. After a couple of months, the third man I had met from Acorn called from LA and thanked me for my help, and said they'd call me if they needed me again. I urged him to give it a little more time, that I thought this was a very good deal for Acorn and for me, and he agreed to continue for the time. I've been sending them monthly invoices every month since, for more than ten years.

The company actually did get a patent, in my name, on the arsenic suggestion I made before I was hired, but no one has yet paid royalties to use it. The three of us published a paper together and I keep writing papers on our work as we go, but Paul and Andreas never seem ready to publish and Acorn doesn't care. Publishing at this point is of no importance to me, but it seems to me that it might be for Paul and Andreas. It's quite remarkable for me to still be employed at 88. I keep paying into Social Security, as well as collecting, and my tax man says I get paid about the highest rate from them that he has seen. It's only a tiny difference but few people work this long.

Occasionally I substituted teaching for other faculty who needed to be away. In connection with doing one of Z. X. Shen's classes I thought of a shortcut for desalinization of seawater. The idea was to lower a bell deep enough in the sea that the water pressure was enough to drive fresh water through a semipermeable membrane into the bell, kept at surface pressure with a tube to the surface. It was easy to calculate that this would occur below some 800 feet depth. The device is equivalent to an 800-foot-deep freshwater well. I took the idea to the Stanford patent office and they would make the application for a patent, which first involved a search for past patents. They quickly found that some fifty years earlier someone obtained a patent for exactly that device. That meant that I couldn't patent it, but it also meant that anyone could

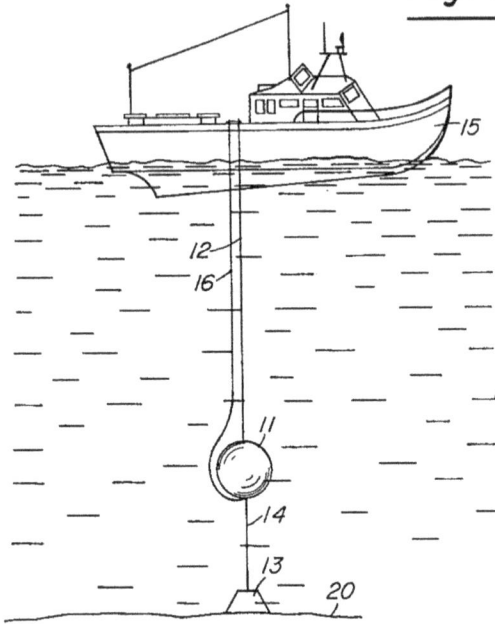

Sketch of a sea-well, from the patent application by F. D. Carpenter in 1962.

use the idea and not have to pay royalties because the patent expired after 20 years.

I really thought it was a good idea and took it upon myself to try to promote it. I read that they were planning to build three desalination plants on the Monterey Peninsula and after a little checking it looked perfect. There was a deep trench coming out from Pajaro Slough into the bay so water 900 feet deep could be found close to the shore. I was able to get in contact with the engineers who were preparing for the facilities. They were interested, but were not ready to consider anything different than the traditional method of pumping it into a factory, building up pressure to drive it through a membrane. When what's left behind gets a little more concentrated it takes higher pressure, and eventually it won't flow and they pump the brine back into the bay, causing concern from the environmentalists. Then they start over. It is sad, but maybe not surprising, that they weren't willing to try something new.

Bill Shockley was a physicist with whom I ended up having a surprisingly large number of interactions. I met him first as a student at my first American Physical Society meeting. At breakfast in the hotel I saw my advisor Fred Seitz sitting by himself and joined him. Later Shockley came and joined us both. He won the Nobel Prize for the invention of the transistor that year, with John Bardeen and Walter Bratain, and soon moved to Palo Alto to set up the Shockley laboratory. He realized that such devices were the way of the future and preferred that he, rather than Bell Labs, should cash in. He managed to hire an extraordinarily able set of young coworkers, but he was so difficult to work with that the group resigned as a unit, joined Fairchild Corporation and founded silicon valley. Shockley was furious, but went ahead and replaced these "traitors" with a set of coworkers from Germany whom he could control. The company did not do well and he ended up taking a professorship in Electrical Engineering at Stanford. His office was at the end of my corridor in the McCullough Building when I came to Stanford.

I would occasionally have short talks with him after I came to Stanford. At one point he got very much interested in a puzzle, originated by Richard Feynman, I believe. It had to do with a superconducting ring, sitting in space, with current flowing in it, with a resulting magnetic field through the ring. As it warmed up, the superconductivity would suddenly disappear, the current would drop to zero, generating a pulsed electric field around the ring. If a positive electric charge had been attached at a point on the ring on one side and a negative charge attached at the opposed diameter, the force from this electric field would be in the same direction on the charges on both sides, and the ring would spontaneously start moving through space. No one believed that such a ring would really spontaneously take off but it was hard to see the error in the argument.

Bill Shockley believed that it *would* take off, that the combination of electric and magnetic fields corresponded to static electromagnetic momentum, which was suddenly converted to dynamic physical momentum. We had at the time at Stanford a small club of physicists called the Maxwell Society which occasionally met in the evening. I invited Bill to give a talk to this group, which included "Pief" Panofsky, a very bright and widely respected physicist at SLAC. Bill gave his talk to Pief, with the rest of us listening in. At the end, Pief sat for a few moments and then said: "There are three places where you may have made your mistake. The first, which I am almost certain is the place, is neglecting relativistic effects, which always need to be considered when magnetic fields are involved." I don't remember the other two. Bill said: "That is very interesting. I will look into those carefully." And that was the end of Bill's belief in the puzzle. It was also approximately the end of the Maxwell Society. Some years later I ran into Panofsky at a reception and told him that story. I was surprised that he had no memory of it.

In a similar vein, Shockley once remarked to me that some people think of the repulsion between two like-charged bodies as a long-range interaction, and others think of the force as resulting from the electric field energy in space arising from the two charges. He said the second was right: if at some moment you made the space electrically conducting, all that field energy would be deposited locally as heat in direct proportion to the squared electric field at that point in space. At the same moment the force on each charge would disappear. I would say that he was exactly right, but I had never thought of it that way.

At some time Shockley got involved in Eugenics and race and the improvement of mankind by selective breeding to improve IQ's. That was generally unpopular, and at the old-boys' lunch, which he sometimes joined, he brought it up and reached for a pamphlet in his brief case. Someone, I think Sandy Dornbusch, said "No, Bill, no pamphlets at lunch!" and that took care of it.

Around that time Lucky and I were in New York for a physics meeting, having a drink with friends in a hotel lobby. I had heard a talk that day by John Bardeen on the invention of the transistor and he had given Shockley complete credit for coming up with the idea and suggesting the experiments. I thought at the time that I wished Shockley were there to hear it. Then Shockley walked by our table in the lobby and I asked him to join us, which he did. He *had* heard Bardeen's talk, but was not much impressed. He started in on Eugenics and Lucky got upset, saying: "You want to kill Candy Linville?" with tears in her eyes. Candy was the blind daughter of a friend of all of us. It's the only time I can remember Shockley being shaken up!

Back at Stanford the black community was very much exercised over Shockley's statements about intelligence and race, and started a move to get him fired from the Stanford faculty. They held a rally on White Plaza and Shockley went to it, wearing a sign with something like "Truth and Free Discussion" written on it, standing some feet from the podium. I'm sure it made the speakers uncomfortable, but they proceeded and didn't ask him to speak. When it was over he wandered off. It's hard to imagine someone putting himself through that.

Some time at the GE Research Laboratory Ivar and a group of others were trying to evaluate the impact on science of different people in the Lab. They were using the Citation Index as a measure. You could look people up and see how many references were made by other people to their work. They were surprised to find Bob Fleisher, who could be thought of as a physical metallurgist, to be the leading one at the Research Lab. For comparison, or normalization, they looked up some famous people from outside the lab, including Shockley. As they expected, Bill had more references than anyone at the Lab, but not that much more, as they would have thought. Then they realized they had misspelled his name as Schockley. Then they looked up the correct Shockley and the list went on and on. We all had a laugh over it and I thought Bill would enjoy the story. When I happened to be sitting next to him at a colloquium, I told him the story, and all he did was make a wry smile and say: "Oh, they misspelled my name?"

More recently there was an International Conference on the Physics of Semiconductors to be held in San Francisco and I was involved in the organization. Shockley hadn't been to such a conference in many years, but I suggested he come to the main social event and dinner. I said he wasn't invited to speak, just to go and probably run into many old friends. He agreed and I took tickets to his house for the event for him and his wife. I told him to let me know if he couldn't come because the organizing committee was paying for them and had a limited budget. He didn't show up, and afterwards when I asked him what happened he said he needed to write a brief for a lawsuit he was involved in and didn't have time. I let it go at that. He died a few months later, and I never told him that we didn't have to pay for their tickets in the end.

I mentioned another interesting character earlier, Brian Josephson, who took me to a college fest in Cambridge and who became famous for the Josephson effect that Bardeen didn't accept at first. One time in Cambridge we were discussing this Josephson Effect and he kept saying it was like a pendulum. It seemed strange to me because physicists always use the term "harmonic oscillator". Then I realized he was talking about a *real* pendulum, which can go over the top, not one with small displacements which is a harmonic oscillator, and I could see that his parallel was an exact one and described the *ac* Josephson Effect as well as the *dc* one. At this same time, Brian was getting interested in psychic phenomena. He wanted to show me an electronic device he had which had psychic powers, but he couldn't find it in his apartment. He even told Ivar Giaever that his guru could levitate himself by concentration. Ivar said: "Now Brian, have you seen him do it?" Brian said no, but he knew he could do it.

Later Brian was visiting the bay area with his wife and contacted me. We invited them out to dinner. He brought a wild older man and his girl friend who were also into psychic things and were probably the reason for the Josephsons being in the area. They arrived at our house before dinner and needed to have time in separate rooms, with absolute quiet, in order to meditate. That was not easy with our four sons in the house. However, we got through it and went out to an Indian restaurant for one crazy dinnertime.

For a long time, Lucky and I had stopped at Antonio's Nut House (picture in Chapter I) for a wine and beer, respectively, if we had been out for dinner. It would be called our "local" and we knew many regulars there, including the bartenders and Tony, the owner. One time I mentioned it to Jim Gibbs, my Anthropology friend from Cornell, and he said: "Oh, that working-class bar at the end of California Avenue." And I guess it is.

Some time Lucky thought maybe we could try a little classier place and we went to Sundance Mining Restaurant and Bar. The drinks cost a little more but the music wasn't so loud and the ambiance was definitely nicer. In those early times, it was usual to smoke in bars and Lucky enjoyed a cigarette with her wine. I was no longer smoking at the time.

Then the law was passed outlawing smoking in bars and of course Sundance followed the rules, but Tony didn't. We switched back to Antonio's. I was quite happy with that. Somehow Antonio's makes me feel comfortable and at home. Eventually, Tony had to follow the rules and it was necessary to go outside for a smoke, but we didn't switch back to Sundance. Antonio's has been our local since then.

Our friend Siegie comes regularly to Antonio's and is our age. However he grew up in Germany so when I joined scouts he joined Hitler's Youth. He can still remember the wonderful smell of his wool uniform when his mother bought it for him. He remembers his family's doctor who was Jewish. The doctor and his family disappeared during this time.

I remember one occasion when I was sitting at the bar next to an Electrical Engineering graduate student who seemed terribly distressed. He had been talking to a younger physics student about quantum mechanics at the bar and found that he had been quite wrong about some aspect of his understanding. He felt he could never be successful in engineering if he had been so wrong about quantum mechanics. I assured him that that is one of the ways you really learn, when you find you've been wrong. He seemed somewhat comforted and I thought: this is a rare bar where the thing that someone is so distressed about is something as arcane as quantum mechanics. More recently I recall a young man with very long hair who also spends time in Madagaskar studying lemurs. If this is a workingman's bar, it is one with an unusual clientele. Maybe it is duplicated in some place like New Haven, Connecticut.

On a number of occasions we have taken visitors from Europe to Antonio's and they always seem to find it very different and memorable. In about 2016, Tony died and I went to his funeral. Kelly, one of the earlier bartenders, and Tony's son continue operating the business and it seems much the same to me. Lucky thinks it has changed and resists going but occasionally we drop in and sometimes I have gone alone.

I seem to be invited to meetings about as often after retirement as before. On one European trip, Lucky and I went to Aarhus, in northern Denmark, invited by Niels Christiansen who had met us at the beginning of our first stay in Stuttgart. He had since moved on to this city in Jutland. In the evening after dinner Lucky and I went to a bar. In came Per Arnoldi, whom we didn't know, but soon did. He was a very notable maker of posters in Copenhagen who had posters up in all the railway stations. He had been interviewed on television in Aarhus and was all hyped up and came to the bar with a girl from the station. We all got on well, had a pleasant evening and kept in touch.

The next time we were in Copenhagen we looked him up and some time later he was visiting Stanford in connection with some of his art being shown and contacted us. I took him to lunch at the Faculty Club and after that he came to my office. I had a brand new Macintosh, the original lunch-box style, with Macpaint on it. He had never done computer graphics and gave it a try. He wanted to print five copies, which we did. He numbered and signed them, and gave me #1, which I still have in my office, though suffering from the time when coffee was spilled on it. He had also wanted it erased on the machine with no possibility of recovering it, which we did. The mind of an artist seems to proceed differently from that of a physicist.

Per Arnoldi's first graphic print, made in my office at Stanford.

Jan Zaanen, whom we had known from our stays in Stuttgart, and his German wife Krista were now settled in Leiden, the Netherlands, and we visited them there once on a European trip. They came to Stanford twice for sabbaticals and for some shorter stays. Lucky and I would see them frequently when they were here. He sometimes came to the old boys' lunch with me, and was referred to as my Dutch friend.

Jan and I did not always see eye-to-eye on research. I had been exploring something in the theory of magnetism called Heisenberg exchange. I found that the traditional approach, which was almost universally used, missed half of the contribution. I was quite excited about it and mentioned the finding to Steve Kivelson, a very able Stanford theorist. He said "Obviously!" and made gestures with his hands that clearly made the same point. Thus I learned, (i) Sometimes able people know a lot that they don't bother to mention. (ii) I shouldn't make an issue of this discovery. Jan Zaanen was one of the people who, jointly with a student, had explicitly used this traditional approach in a publication where his result nevertheless agreed with experiment within 10%. I wasn't surprised because I think that often students are pushed to make their theory agree with experiment, and I pointed it out to Jan. He exploded, "No, you don't understand! You have to read the many-body literature which proves that our approach is correct." I said, "No, Jan, I don't have to read all the literature. It's a simple point and you can take it or leave it." Sometime later he said: "Well, Walt, a factor of two here or there doesn't really matter. If we made a mistake of a factor of two, Phil Anderson made the same mistake in the work for which he won the Nobel prize." That was true, but the difference was that Phil had opened up a whole new field with his approach and the factor of two *was* incidental. I thought Jan's student had just fudged a calculated result to agree with experiment, but I didn't mention that.

Many retired people do their traveling with organized cruises. That didn't come up for us earlier because we had so many occasions to travel in connection with my work. I could make travel arrangements to match some meeting and then we could expand the trip with stops on the way or the way back. Two of these did involve a cruise ship. One was a visit to St. Petersburg after a stop in Sweden. A physicist in Sweden arranged a cruise from Sweden to St. Petersburg for the two of us on a Russian ship. We had first class accommodations, windows facing forward and a private table for meals but that wasn't much fun. When we arrived we had to wait for an hour to disembark while some communist dignitaries were looked after, and then we ended in a dark lot. Fortunately two young Russian physicists were there to look out for us. We had reservations at a Best Western St. Petersburg but the cab took us to a dark building with no signs. One of the young Russians explored and found indeed that deep inside was the hotel and it worked out well. I went to the Institute the next day and we went to dinner in the apartment of my old friend Ija Ipatova, and Igor Abarenkov was also there.

When we went to an International Conference on the Physics of Semiconductors in Jerusalem, we arranged a cruise on the Nile to see the Valley of the Kings and other high spots. It was the year there had been a massacre of tourists at the Temple of Hotshepsut so there was lots of security. It was a small ship but had some aspects of a usual cruise. There was a pool on the deck, we had meals in a dining hall, and each day there was a bus available to take us to the site for that stop.

The family at the beach recently. From the left John and his partner, KK, Bill in back and Rick's Linda in front, Rick, Walt, grandson Ace in back, Lucky in front and then Bob. Missing are Bill's Geng and Bill's sons, Luka and Kasper, as well as Rick and Linda's Mark and Jennifer.

Finally we did a more standard cruise. I was interested to see the Panama Canal, having read about its construction. From Puerto Rico we took a sizable ship and made a stop in Caracas, Venezuela, where we shared a cab with another couple from the cruise to see the city. We went through the canal and made several stops along the west coast on the way to Acapulco. We had a balcony from our room so the viewing of the canal was good. On board we had drinks in the small bar before dinner every night, which wasn't heavily attended, and enjoyed the few patrons whom we got to know. We very much enjoyed the experience and talked about it frequently afterward, but never had the urge to do another cruise. Lucky liked the idea of a river cruise, but we never did one. I think we're not cruise people.

There was some talk in the physics community about dark matter and dark energy in the universe. I considered it hype by people trying to get astrophysics money from the government. It even occurred to me that the observations which suggested dark matter could be explained by a slight increase in the strength of gravity at galactic distances, and the observations suggesting dark energy could be explained by a slight *decrease* in the strength of gravity at *intergalactic* distances. When at a physics dinner a graduate student was enthusiastically talking about dark matter, I suggested this explanation. He said: "Oh, you're one of the MOND guys." It was a jolt to be reduced to a category and I asked what MOND stands for, "Modified Neutonian Dynamics." They say there's nothing new under the sun.

It reminded me of an earlier time when I decided to get to the beach where the family was staying by public transportation. I walked to the California Avenue station and took the train to San Jose, and then to the bus stop. I climbed on the bus to wait for the departure for Santa Cruz. There was a radio playing on the bus, and when the driver got on he commented: "What a pain to have to listen to this stuff all day." I'd been listening and said: "Well at least you get to learn about the intergalactic dark matter," which had been discussed on the program. He said "Well, yes, but what they really mean is interstellar, not intergalactic, dark matter. As you know, it's 25,000 light years to the nearest galaxy and only four light years to the nearest star, Alpha Centauri." I said: "Wow! I really hit on your hobby!" He said, no, he had no interest in it but he remembers everything he ever hears. I said well, maybe you can answer a question we were asking the other day: what's the order of the planets. He started with Mercury and gave the distance and number of moons on each planet, all the way to Neptune. It's remarkable who you might meet in unlikely places!

Life for us continues to flow along quite smoothly. We see our California sons and their partners regularly. I make it to lunch with the dwindling number of old boys at the Faculty Club on Fridays. I spend a day a week with Andreas and Paul for Acorn. Lucky and I still have a trip planned to New York City. I'm the only driver, but we make it to our various appointments and parties, and I now have a three-wheel bike that I can use to go to Acorn or the Stanford Faculty Club. I spend time working on Acorn problems, but could make time to put together this memoir.

If I could have picked the time to live my life, I think I would have picked my actual birthdate, within a year or two. That was my first break! I was too young for World War II, and when it came time to graduate from college, my choice was either to go to graduate school or to be drafted, and I chose the former, and am grateful to have done that. It seems unlikely that I would have continued in school without the pressure of the war. I was in graduate school just three years, but then the war was over and I was at GE. When the Vietnam war came I was too old for the draft, but had to worry about my growing sons. They were too young and then the draft ended, so all of us have escaped that experience.

A second break was that solid-state physics was a small world when I entered it. We were just beginning to understand the electronic structure of solids and computers were brand new. Semiconductors were just becoming significant, and the transistor was invented at Bell Labs. The significance of vibrations of crystal lattices was being recognized and even my representation of them in my PhD thesis could be a significant contribution to the expanded understanding.

A big step was Pippard's realizing that for metals a Fermi surface was a real shape that was relevant and he determined that shape for copper.

Before that, the standard procedure was to imagine an ellipsoidal shape and adjust its parameters to accord with some experiment — with different shapes obtained from different experimental properties, and no possibility of coming to a firm conclusion. The big change came from Pippard's recognizing that there was a *real* Fermi surface that could be determined, and all interpretations of properties had to conform to that surface. I knew personally almost everyone who was working in the field of the electronic structure of solids and we felt that we were working together with an important common goal. Since then the number of people in the field has continued to expand, with each person confining himself to a smaller and smaller part of it. It is a vibrant technical activity and I had had the opportunity of joining it at the beginning.

A third break was of course being lucky enough to have the wife and family I had and to enjoy the adventure with them.

Lucky and Walt, recently in New York.

INDEX

.